SOTHIS PRESS

Sothis Press aims to completely reimagine the public domain book with an emphasis on universal access to captivatingly designed print editions. We believe that technology will enable every single book ever published to be perpetually accessible in beautiful print form. Sothis is beauty in ubiquity.

www.sothis.press
1 Waterhouse Square, London, ECIN 2ST

Yellowstone Park Guide

A Practical Hand-
Book, Containing Accurate and Concise
Descriptions of the Entire Park Region, Maps,
Distances, Altitudes, Geyser Time Tables and
All Necessary Information; Profusely Illustrated

A. B. Guptill

SOTHIS PRESS

ISBN 978-1-0368-0735-1

This edition is published by
Sothis Press © 2023. The company
is registered in England and Wales.
The Company Number is 14900389.

Haynes Guide

TO

YELLOWSTONE PARK

A Practical Hand Book

Containing Accurate and Concise Descriptions of the Entire Park Region, Maps, Distances, Altitudes, Geyser Time Tables and all Necessary Information, Profusely Illustrated. : : : : : : : : : : : : : : : : :

By A. B. GUPTILL

Illustrated and Published by F. JAY HAYNES
Official Photographer, Yellowstone Park,
St. Paul, Minn.

THE SHOP BY THE RIVER
THE PIONEER PRESS
SAINT PAUL
1849

CONTENTS

CONTENTS

ILLUSTRATIONS

ILLUSTRATIONS

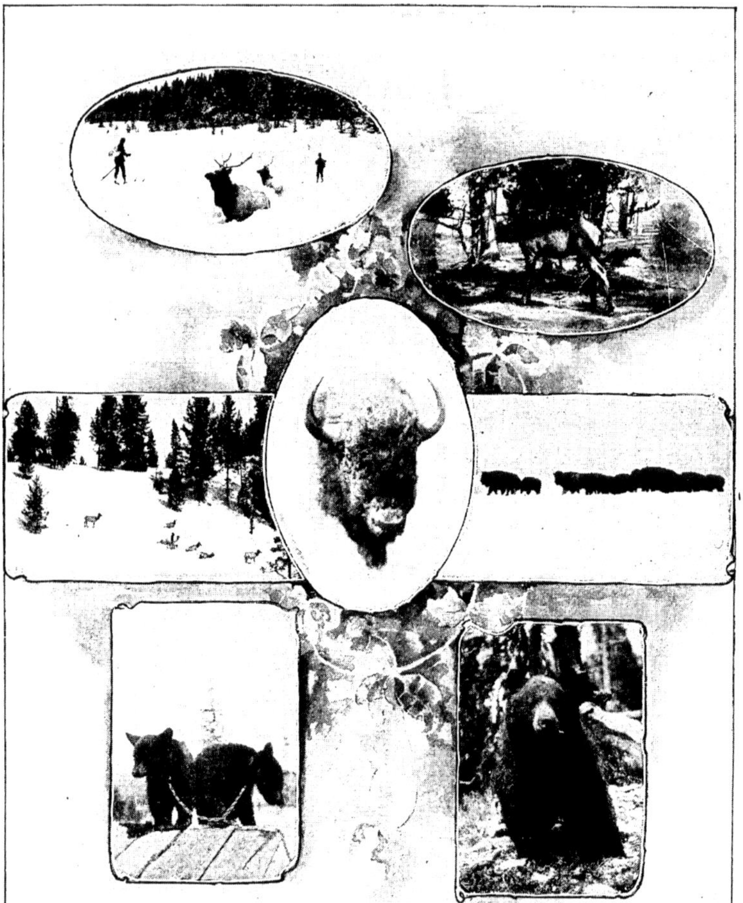

WILD ANIMALS OF THE PARK.

Visitors entering the Park via the O. S. Line or Madison River Entrance, see page 40.

Yellowstone National Park

In the northwest corner of Wyoming, in the heart of the Rocky Mountains, is located Yellowstone National Park. Its boundaries overlap a few miles into Montana on the north and Idaho and Montana on the west. The reservation is about sixty-five miles east and west and seventy-five miles north and south. No valley within its limits has an elevation of less than 6,000 feet, while many of the mountain peaks within and adjacent to the Park rise from 10,000 to 14,000 feet above sea level.

Yellowstone Lake, 15x20 miles in size, is the largest body of water in North America at this altitude (8,000 feet). Three of the largest rivers in the United States —the Missouri, Yellowstone and Columbia—have their source in the Yellowstone Park. The geysers of this region outclass anything of the kind in the known world. There are over fifty that throw a column of hot water 30 to 250 feet in the air at intervals of one minute to fourteen days. The Grand Canyon of the Yellowstone, ten miles long, with an average depth of 1,200 feet, is acknowledged by travelers to be the most brilliantly colored landscape in existence. The Mammoth Hot Springs are the only colored terrace-building hot springs known that have such beauty and magnitude. Cliffs of volcanic glass, unsurpassed waterfalls, mountains of petrifactions, charming valleys, hills of brimstone, perpetual snow-clad peaks, interspersed with thousands of natural curiosities, fittingly characterize this as the wonderland of the world. Modern hotels have been constructed throughout the Park, conveniently located near these objects of interest. Substantial roads and bridges have been built leading to all the chief attractions—steamers have been placed on the lakes, mountain streams have been stocked with rare species of the finny tribe, military posts have been established, railroads have approached the Park boundary—all for the pleasure, comfort, protection and enjoyment of the people.

THE CONTINENTAL DIVIDE.

This great range of mountains, extending from Mexico to Canada through the western portion of the United States, is the water-shed of the continent and a salient feature of Yellowstone Park—entering near the southeast corner and extending northwest between the headwaters of the Yellowstone River on its eastern slope, and Snake River on the western, separating Yellowstone Lake from Shoshone Lake by less than a dozen miles. On the summit of the range, a few miles south of Yellowstone Lake, is *Two-Ocean Pond*, a small lake with two outlets, one into the Atlantic through the Yellowstone and Missouri Rivers, the other into the Pacific through Snake River, a branch of the Columbia. The divide passes within a few miles of the upper geyser basin and makes its exit from the Park about midway of the western boundary; the general direction after leaving the Park is westerly for about fifty miles and forming the dividing line between the states of Idaho and Montana.

EMIGRANT PEAK, MONTANA.

LIVINGSTON TO MAMMOTH HOT SPRINGS.

Livingston, Montana, is on the main line of the Northern Pacific Railway, midway between Lake Superior and the Pacific Coast. From this point a branch of the road extends fifty-four miles south to Gardiner, Montana, at the northern boundary of the Park, following the valley of the Yellowstone the entire distance. Livingston is located at the base of the Snowy and Belt ranges—spurs of the Rocky Mountains. Three miles from the town the Park branch passes through the first canyon of the Yellowstone or Gate of the Mountains, which forms a natural entrance into the Upper Yellowstone Valley. This canyon is nearly a mile long and just wide enough to comfortably admit the road and river, the mountain walls rising some 2,000 feet perpendicularly on either side. Passing through the first canyon, Paradise Valley is immediately entered; it extends from the mouth of the canyon some thirty miles up the river and is from seven to twelve miles wide. This valley has been settled by ranchmen for several years; it is very fertile, easily irrigated, and well adapted to stock raising. On the east side of the valley, a very picturesque mountain range is seen, constantly changing as the train moves along; the hills on the west are not as abrupt, but are very interesting from a geological point of view. Emigrant Peak (elevation 10,629 feet and some 6,000 feet above the valley) is a prominent mountain, located near the south end of Paradise Valley.

The Second, or Yankee Jim, Canyon.—Forty miles from Livingston the Park branch passes through the second or "Yankee Jim" canyon of the Yellowstone. For several years prior to the building of the railroad, Mr. James George, an early prospector and an enterprising individual, having constructed a wagon road through the canyon, enjoyed a lucrative business in collecting toll from each visitor to the Park.

Yankee Jim is still living on his ranch at the south end of the canyon and has many guests during the summer months, who find in the swift waters of the river the best of trout fishing—equal to any of the many excellent fishing stations along the Yellowstone. The second canyon is far more picturesque than the first; the mountain walls rise higher, and the gorge, within which the river is compressed, is scarcely a hundred feet wide. An excellent view is had from the cars as the train winds slowly through the gorge.

Cinnabar Mountain, a conspicuous landmark on the Upper Yellowstone, is about three miles from the Park line. As the train passes along its base, one has a glimpse of the "Devil's Slide," composed of a reddish colored mineral resembling Cinnabar, and extending up the mountain some 2,000 feet.

Electric Peak, the highest mountain in this vicinity, is directly on the northern boundary of the reserve, the line passing over it. It is accessible from the north, a trail leading nearly to the summit, and can also be ascended from the south, but saddle animals cannot be taken as near the summit as by the former route.

YELLOWSTONE NATIONAL PARK.

Gardiner, Mont., the terminus of the Park branch railroad, is a supply point for the mining camps in the vicinity, and an outfitting place for hunting and camping parties. The unique log depot constructed in 1903, and the imposing stone Arch on the northern boundary, dedicated by President Roosevelt, April 24, 1903, add much to the attractiveness of this entrance to Yellowstone Park. Tourists are conveyed from the station to the Mammoth Hot Springs Hotel, five miles distant, in six-horse tally-ho coaches. The stage road soon enters the Gardiner Canyon, following along the beautiful Gardiner

THE ARCH, AT NORTHERN ENTRANCE.

River, over one of the most attractive drives in the Park, passing beneath Eagle Nest Rock where the canyon cliffs are fully 1,500 feet above the roadway.

MAMMOTH HOT SPRINGS.

The terrace-building hot springs of Yellowstone Park are the most remarkable development of thermal action to be found, occupying several acres, with many distinct terraces and springs. The present active portion of the hot springs is located on the eastern slope of Terrace Mountain, from 50 to 300 feet above the plateau, upon which the Mammoth Hotel, Fort Yellowstone and other buildings are constructed. This plateau was formerly an active portion of the hot springs. It contains many extinct craters and openings, some of them leading into immense caves, but the presence of obnoxious fumes and gases makes it impossible to explore them.

VIEWING THE TERRACES.

It requires fully two hours to visit all the prominent springs and terraces. Still the construction of the new road to Golden Gate, which passes along the base of "Jupiter Terrace," gives one, while riding past, an excellent idea of the form and coloring of one of the largest terraces. To thoroughly inspect the beauties of these wonderful springs, one must do considerable walking, whether he takes a carriage to the top of the formation, or walks the entire distance. The "Formation party" starts from Mammoth Hotel about 2:30 p. m., in charge of a competent Y. P. A. guide, for a stroll of two hours among these wonders. Do not bur-

LIBERTY CAP AND MAMMOTH HOTEL.

den yourself with heavy wraps, as it is much warmer on the formation than in the shade of the hotel veranda.

Liberty Cap, an extinct hot spring cone, standing at the foot of Terrace Mountain, near the road, is fifty-two feet high and twenty feet in diameter at its base. It is formed of over-lapping layers of deposit, evidently having been built up by the overflow of water through the orifice in the top. Scientists have been unable to decide whether it was built up independently or formed by the action of the elements wearing away the soft material surrounding it.

The Devil's Thumb, a cone of similar structure, but smaller, is located some 200 feet west of Liberty Cap, partially imbedded in the hillside. The path leading to the formation past the Devil's Thumb is generally taken when returning, the one for the ascent branching off the main road a short distance south of Liberty Cap.

Minerva Terrace is a mass of deposit forty feet in height, covering an area of nearly three-fourths of an acre, with a hot spring on the summit some twenty feet in diameter, the temperature of which is, at the edge, 154 degrees Fahrenheit. The constant changing of the overflow and the intermittent character of the spring make it impossible to predict, a season in advance, which will be the active side of the terrace, or whether it will be active at all. At times the spring disappears entirely, and the terrace remains inactive and uninteresting in consequence for months. The change in overflow, when the spring is active, is accounted for by the rapid deposition of carbonate of lime, which forces the water eventually over the entire surface.

MINERVA TERRACE.

The quantity of water overflowing is very small compared with the amount of deposit, which, under favorable circumstances, is about one-sixteenth of an inch in four days. Articles of iron, glass, or any hard substance placed where the water can run over them, are soon coated with a crystal-white deposit. During periods of activity, basins or pools, fringed with stalactitic masses, line the east side of the terrace, presenting the most delicate coloring, from the lightest cream at the top to the deepest shades of yellow at the base, the predominating color being bright orange, each pool or basin being filled with transparent blue water. The elevation (some twenty feet high) immediately back of the terrace is an excellent point from which to obtain a good view of the interior of the spring. This terrace is about seventy-five feet above the level at the base of Liberty Cap, while the main portion of the Hot Springs is on the mound some ninety feet higher.

Jupiter Terrace, the largest of the group, extends some 2,000 feet along the edge of the high mound of brilliantly colored deposit immediately south of Minerva Terrace. A climb of about 100 feet, up quite a steep trail, is necessary to reach the summit. The two large springs of boiling water, fully 100 feet in diameter, supply the main terrace, seen from the driveway, as well as the beautiful "Pulpit" Terraces beneath, on the eastern slope. From the summit of Jupiter Terrace one sees, in a northwesterly direction, a vividly colored mass, with its background of dark green pine trees. This is "Cleopatra" Terrace.

From the prominence west of Jupiter Terrace, under

which is located Cupid's Cave, an excellent general view may be had. The path leading west from Cupid's Cave passes along the summit of Narrow Gauge Terrace, which terminates at the hill, where the main path leading to the hotel is intersected.

Narrow Gauge Terrace.—A fissure ridge 300 feet long. When active is filled with miniature geysers and springs which deposit the most brilliant coloring.

Orange Geyser, on the terrace above Narrow Gauge, is greatly admired by all visitors. It consists of an oblong mound of deposit some twenty feet high and about thirty feet in diameter. The active little geyser on its summit and the brilliant coloring are its chief attractions.

Bath Lake is a few hundred feet south of Orange Geyser, separated by a timber-covered ridge of ancient deposit that nearly surrounds the lake. There is no visible outlet to Bath Lake; however, the uniform temperature of the water at all seasons of the year is one of the mysteries of this region.

Devil's Kitchen is the crater of an extinct hot spring. It can be entered with safety. Through a small opening some six or eight feet in diameter you descend a ladder into the kitchen. The peculiar damp and heated atmosphere of the interior produces a queer sensation and the desire to seek fresh air at once comes over the visitor. When the Devil's Kitchen was first explored (in 1881) numerous bones of wild animals were found in the cave and it was alive with the flying bat.

ANGEL TERRACE.

Angel Terrace.—In visiting the Hot Springs by carriage the order of the trip is reversed and Angel Terrace is visited first. It is situated on the extreme southern part of the formation and about 1,000 feet south of Orange Geyser. For delicacy of coloring and beauty of form there is nothing to compare with it among these wonders of thermal action.

PARK PROTECTION.

Fort Yellowstone.—At the Mammoth Hot Springs is a two-company United States cavalry post, the commanding officer being superintendent of the Park. During the summer months cantonments are scattered throughout the Park, their duties being to protect the various objects of interest from vandalism, see that no specimens are removed, that no poachers enter the Park, and prevent the spreading of camp fires. The reservation is entirely free from questionable characters owing to the rigid enforcement of the orders of the commanding officer. Several scouts are employed by the government, who roam over the entire area; they look after the game chiefly, their occupation being similar to that of a detective. Much credit is due the military and scouts for their efficient efforts in protecting the Park. The United States Commissioner's Building is a substantial stone structure, modern throughout, combining residence, court room and jail. The rules providing for the government of the reserve will be found on back pages of this hand-book.

Improvements in the Park the past season have been many. A new Hotel has been erected at the Upper Basin and a Lunch Station at Thumb Bay, eliminating the canvas hotels. The United States Engineer Road Department has completed its system for sprinkling, daily, about one hundred miles of the Park roads, constructed a magnificent driveway over the Yellowstone near the Canyon, and replaced many of the old wooden bridges with steel structures. A Government Weather Bureau Station has also been established near Fort Yellowstone.

FORT YELLOWSTONE.

TOUR OF THE PARK.

SOUTH FROM MAMMOTH HOT SPRINGS.

Silver Gate and the Hoodoos.—The driveway between the springs and Golden Gate ascends the mountain with such easy grades and graceful curves that one does not realize that a thousand feet elevation is gained in less than three miles. This road passes through the "Hoodoos," a wild, strange region heretofore inaccessible. Many theories are advanced as to the origin of the "Hoodoos." The most plausible is that the immense quantity of deposit or formation seen lower down the valley, even as far as Gardiner River, two miles distant, was carried there in solution by the

SILVER GATE.

hot waters of Mammoth Springs, thus leaving honey-combed caves beneath, and the present "Hoodoo" region is simply where the mountain has caved in, filling the cavern below. They cover an area of about one square mile, and are located nearer Golden Gate Canyon than the Mammoth Springs. In the midst of the "Hoodoos" the road makes an abrupt turn, passing between great blocks of limestone that rise abruptly fully seventy-five feet, to which is applied the very appropriate name, "Silver Gate."

Golden Gate.—Four miles from Mammoth Hot Springs is one of the most picturesque points in the Park. It is a rugged pass between the base of the lofty elevations of Bunsen Peak and the southern extremity of Terrace Mountain, through which flows the west

GOLDEN GATE CANYON AND VIADUCT.

branch of Gardiner River. The sides of these rocky walls, which rise 200 to 300 feet above the roadway, are covered with a yellow moss, suggesting the appropriate name the pass now bears. The pillar at the east entrance, some twelve feet high, was originally a part of the canyon wall. The construction of this roadway and viaduct, scarce a mile in length, was the most expensive and difficult piece of road building yet encountered by the government engineers.

Rustic Falls, occupying a conspicuous position at the west end of Golden Gate, adds a charm to this beautiful spot, and when seen in the early part of the season is especially fine. The stream is fed by mountain snows and springs along the base of the hills a mile or so beyond. The fall is some sixty feet, over a series of shallow basins worn into the dark, moss-covered ledge, disappearing underneath the surplus of rock deposited in the canyon from the construction of the roadway. The view obtained of Golden Gate upon the return trip is equally interesting.

Swan Lake Basin.—A pleasant surprise awaits the visitor immediately beyond Golden Gate, in Swan Lake Basin, it being quite unlike the region just traversed, and one of the many typical mountain prairies hemmed in by snow-clad peaks found throughout the Park. Evidences of old Indian camps are seen in many places, and during the fall and winter it is inhabited by hundreds of elk and deer. The magnificent range to the west is the Gallatin Mountains, among which are Bell Peak, Quadrant Mountain and Mt. Holmes; the last named having an elevation of 10,578 feet. Vast fields

ELECTRIC PEAK.

of perpetual snow are in sight throughout the summer. To the north about eight miles is Electric Peak, the highest mountain in the northern part of the Park, whose summit is 11,125 feet above sea level, deriving its name from the fact that a great deposit of minerals renders the workings of the surveyor's transit impossible when on the mountain. The peculiar electrical display from its rugged peaks during a thunder storm is a sight witnessed by only a favored few. The drive continues south through Swan Lake Basin, nearly upon the same level, to Norris, crossing Gardiner and Willow Creeks, the two forming the middle Gardiner River.

The Apollinaris Spring is on the east side of the road near the ten-mile post—a delicious spring of natural Apollinaris water, as refreshing as the genuine article of commerce.

Obsidian Cliff.—This bold escarpment of volcanic glass is twelve miles south of Mammoth Hot Springs. The roadway passes along its base for 1,000 feet between it and Beaver Lake. The vertical columns of pentagonal-shaped blocks of obsidian, rising some 250 feet above the road, present a glistening, mirror-like effect when illumined by the sun's rays. The greater part of this mineral glass is jet black and quite opaque, with traces of similar formation variegated with streaks of red and yellow. The construction of the roadway along its base was accomplished in a novel manner and with considerable difficulty ; the use of blasting powder being out of the question, great fires were built around the huge blocks of glass, which, when expanded, were suddenly cooled by dashing water upon them, resulting in shattering the blocks into small fragments. This process made possible the construction of this really wonderful roadway, probably the only piece of glass road in the world. There being no other exposed ridge of obsidian in the Rocky Mountains, and this material being more desirable than flint for the manufacture of arrow heads, it was once a famous resort for all tribes of Indians, who congregated here in great numbers. Obsidian Cliff was "neutral ground" to all the Rocky Mountain Indians, and undoubtedly as sacred to the various hostile tribes as the far-famed Pipestone country of Minnesota. Chips

OBSIDIAN CLIFF

of obsidian and specimens of partly finished arrow heads are found throughout the Park, generally at places occupied by the Indians as summer camps.

Beaver Lake.—The roadway continues along the east side of Beaver Lake, which is about one mile long and a quarter of a mile wide. Several beaver dams are constructed across the lake, forming a series of artificial obstructions, each having a fall of from two to four feet. A beaver house, still inhabited, is located near the west shore of the lake. Since the rigid enforcement of the Park regulations regarding the killing of game, Beaver Lake is becoming alive with numerous water fowl, the passing carriages not seeming to alarm them. The reflection of the pine-clad hills among the dense growth of pond lilies which line its shores, adds to the beauties of this lake. The drive from Obsidian Cliff to Norris, though not of especial interest, is over one of the natural "passes" between the headwaters of the branches of the Yellowstone and Missouri rivers, but the ascent of the divide is so gentle, it is impossible to know when it is passed. Twin Lakes, Mineral Lake and Frying Pan are the attractions between Beaver Lake and Norris.

Norris Geyser Basin.—Many tourists prefer leaving the hotel soon after lunch in advance of their coaches, which overtake them near the Monarch Geyser, the walk through the basin allowing a more satisfactory inspection than possible to obtain in any other way. The Yellowstone Park Association guide at Norris accompanies all parties through the basin and sees them safely aboard the coaches.

CONSTANT GEYSER, NORRIS BASIN.

This region, called the Gibbon Geyser Basin in Dr. Hayden's report, was discovered in 1875 by Colonel P. W. Norris, then superintendent of the Park. Since 1881 it has been called Norris Geyser Basin, which name it is quite likely to retain. It covers an area of six square miles, and is one of the most interesting portions of the Park from a geological standpoint, from the fact of its being one of the highest geyser basins in the Park, and many of its active geysers being of quite recent origin.

While the geysers of Norris Basin do not compare in point of eruptive violence with those of the Upper Basin (thirty miles south), they are of great interest to travelers never before witnessing this strange freak of nature; hence it is better to inspect them when first passing through, as they appear insignificant upon the return trip.

The road follows along the ridge on the east side of the basin, affording a commanding view of the surroundings. The first impression one gets, especially upon a cool day when the steam is visible, is that he is entering a manufacturing locality; the terrible noise and rumbling, the hissing of escaping steam and very unpleasant odors excite a feeling of natural but unnecessary caution, as the roadway and numerous paths leading through the basin can be followed with impunity. The Congress, Constant, Black Growler, Monarch, New Crater and Emerald Pool comprise the chief attractions of Norris Basin, while many beautifully formed and delicately tinted springs contribute to the visitor's enchantment.

The flat or valley to the southwest is filled with numerous openings, the water in many being clear and transparent, and in others of a milky hue, constantly boiling and splashing, many of the vents sending forth a disagreeable, sulphurous odor. This section can be visited if great care is exercised; many of the craters, being but thinly crusted, are insufficient to support a person's weight.

The Congress.—The first sight that attracts the visitor is this immense boiling spring, in close proximity to the road, on the left as you enter the basin. It is the largest spring of its class found in the Geyser Basins and is rapidly approaching a geyser. Its pale blue water is in a state of violent agitation, with occasional demonstrations that force the water fifteen or twenty feet above the rim of the crater; the diameter of the same is fully forty feet. For several years there existed near the Congress the "Steam Vent," one of the features of this basin. It consisted merely of an opening in the rocks from which a great quantity of steam was constantly escaping; the roaring of the same could be heard for miles. During the winter of 1893 the "Steam Vent" ceased and the Congress appeared. The first eruptions were of great force and completely blockaded the road with masses of earth and formation.

Constant.—The little geyser at the south end of the flat is the "Constant" or "Minute Man." It has an eruption every sixty seconds, with only a slight variation; the pool is twenty-four feet in diameter, filled with water of crystal clearness. The absence of a cone or deposit surrounding the geysers in this basin, such as is seen around the geysers in the upper basin, tends to substantiate the theory of scientists as to the age of this locality. During an eruption of the "Minute Man" jets of water are thrown forty feet in the air, while the main body is lifted scarcely thirty feet. The overflow is not large, as most of the water returns into the crater.

The Black Growler, with a chimney-shaped opening, is located quite near the road, at the head of a gulch leading from the plateau. Very little water is thrown out, while a large quantity of steam is constantly escaping, producing a peculiar sound. The deposit surrounding the edge of the crater is, at times, of inky blackness, evidently the origin of its name. The water is not clear and has a strong odor of sulphur, which is probably the cause of its turbidity. The geyser, a few feet to the north, known as the "Hurricane," is similar to the Black Growler in the character of its eruption, which is continuous and destructive to surrounding vegetation and foliage.

Emerald Pool.—The path leading to Emerald Pool and the New Crater branches off from the carriage road a short distance south of the Black Growler. Emerald Pool is somewhat concealed in the timber and is a handsome emerald-tinted spring, 40x50 feet in size.

The sulphur-lined basin with coral walls, most beautifully shaped, can be seen to an appalling depth. It is one of the many quiet springs, slightly overflowing. The water is quite hot, having a temperature of 186 degrees Fahrenheit at its edge.

The New Crater.—This geyser is about 500 feet southeast of Emerald Pool, surrounded by huge blocks of dark yellow rock. It came into prominence during the fall of 1891, when quite a commotion, not unlike an earthquake, was observed. When it burst forth a great volume of water was forced out, flooding the ravine leading to the valley below. Since then it has settled down to ordinary eruptions, about every thirty minutes. The rock-covered crater prevents the discharge attaining any great height. Indications are, however, that it will soon be classed among the large geysers.

NEW CRATER GEYSER, NORRIS BASIN.

The Monarch Geyser, the king of geysers in Norris Basin, is situated at the base of the hill, nearly surrounded by a bluff of brilliantly colored rocks, upon the level of the plateau about 1,000 feet east of the roadway. The crater consists of two oblong openings, the larger of which is twenty feet long and three feet wide. Eruptions of the Monarch occur without warning and consist of a series of explosions, frequently more than a dozen, in which columns of water are thrown 100 feet high. The intervals of eruptions are ordinarily about twelve hours. The Fearless, Vixen, and Steamboat are geysers of minor importance, but are well worth a visit, as well as numerous other springs and pools yet unnamed.

Tourists that have walked from the hotel through the basin, meet their conveyance near the Monarch, where a loading platform and resting place have been constructed.

Three miles from Norris Basin the road enters Elk Park, a beautiful valley surrounded by heavily timbered hills and mountains, the Gibbon River quietly winding through it.

A recently constructed driveway along Gibbon River, from Elk Park into Gibbon Meadows, not only avoids a dangerous hill but opens up a very beautiful bit of Gibbon River scenery, conspicuous among the several attractions being the two *chocolate springs*, unlike anything in the Park. One is situated near at hand, between the road and the river, the other being on the opposite bank.

Gibbon Paint Pots.—Half a mile east of the entrance to Gibbon Canyon, surrounded by a dense growth of pine timber, are located these remarkable paint pots, a carriage driveway connecting them with the main road. They consist of numerous openings in the highly colored clay, and are intensely curious, their brilliant coloring and fantastic shapes being the admiration of all. The greater part of the hot springs are at the base of the hill, while the most beautiful paint pot is some fifty feet up the hillside. This, the main attraction, has a funnel-shaped crater with walls of finely ground clay extending about six feet high ; each puff of steam, through the thick, pasty material in the bottom of the crater, molds a perfect rose in full bloom, soon to be replaced by one equally as handsome; visitors should avoid leaving the regular paths, as the treacherous character of this formation renders it quite unsafe.

Gibbon Canyon.—This rugged mountain pass affords the only fairly easy means of exit from Norris Geyser Basin to the valley of the Firehole. The roadway enters the canyon on the east side of the Gibbon River, and follows the latter's course as nearly as practicable for three or four miles, shadowed by precipitous cliffs, at places some 2,000 feet in height. At the northern entrance to the canyon, on the opposite side of the river from the road, a trail leads to the summit of Mt. Schurz, upon which is located Monument Geyser Basin, at an altitude of 1,000 feet above Gibbon River. Interesting as this "basin" unquestionably is, its difficult access, together with the time necessarily consumed in climbing and descending the somewhat steep trail, is, unless to one inclined to scientific observation, scarcely compensated by even the closest scrutiny of the dozen or so crumbling geyser cones—some of them steaming and rumbling, others apparently extinct—which constitute the sum total of attractiveness and give to the locality a distinctive name. Proceeding along the pass, the numerous little puffs of steam arising from either bank of the river, near the water's surface, need no watchful guide to apprise the passerby of countless hot springs with which the gorge abounds. Many of these springs are curious and interesting, and all can be sufficiently observed as one passes leisurely along, without stopping to examine each in detail. One of them, however, Beryl Spring, is rather more than usually attractive, and deserves, as it seldom fails to receive, somewhat of particular notice. The largest boiling spring in the canyon, being some fifteen feet in diameter, is located close by the roadside, about a mile from the entrance to the canyon and can be readily viewed from a passing carriage; the violent boiling of its surface, coupled with the noisy hiss of escaping steam, while lending something of nervous apprehension to the feelings of the traveler, strangely enough possesses no terrors for the stage-horse, although the constant overflow of scalding water from the edges of its basin-like rim pours across the

GIBBON CANYON.

roadway itself. The road, throughout the canyon's entire length, could hardly have been better constructed to afford a more complete and thorough inspection of the wild beauty of rock and glen, and, as it nears the southern exit from the pass, permits a good view of one of the many charming cataracts of this region.

Gibbon Falls, whose waters, tumbling in a foamy torrent down a series of steep cascades on one side of a bold, rocky ledge, and on the other streaming in a thin, shining ribbon of silvery spray from a height of something over eighty feet, fittingly conclude the attractions of Gibbon Canyon.

After leaving the falls the road passes for a distance of three or four miles over a succession of pine-clad terraces until it reaches the valley of the Firehole River. This portion of the drive is greatly admired, from the fact that the recent improved methods of government road building, in which all timber and rubbish have been removed instead of placed in unsightly piles along the roadside, gives the same something of an ideal park-like appearance. The Gibbon and Firehole Rivers unite in forming the Madison River, one of the three principal sources of the Missouri.

The first glimpse of the Firehole is at the cascades near the junction of the road from the western entrance, about two miles from the confluence of the Gibbon and Firehole. Below the cascades the river is confined in a narrow gorge until it reaches the Falls of the Firehole, located a short distance above its union with the Gibbon. After passing the cascades—where a few moments' halt is generally made—the driveway continues for some distance along the banks of the stream. The graceful curves and pine-clad hills extending to the water's edge, make this one of the many charming

FIREHOLE RIVER AND CANYON

spots of the reservation. The road crosses by a ford the east fork of the Firehole River (Nez Perces Creek), near the summer cantonment of a troop of United States cavalry stationed here for the better protection of this part of the Park, and continues on for a mile or so to the Fountain Hotel, Lower Geyser Basin, one of the best hostelries of the Park—the end of the first day's journey.

THE WESTERN ENTRANCE TO THE PARK.

Idaho Falls to Lower Geyser Basin.

The construction of the railroad from Idaho Falls, Idaho to Yellowstone, Mont. at the Western Entrance to Yellowstone Park, less than twenty miles from the Lower Geyser Basin, has opened a new route that replaces the one from Monida, Mont., a station on the Oregon Short Line sixty miles west of the Park, thereby eliminating the days staging outside of the reservation and materially reducing the time necessary for making the Park trip.

A GLIMPSE OF THE ROCKY MOUNTAINS.

HENRY LAKE, IDAHO.

For several years the branch from Idaho Falls has had its terminus at St. Anthony, Idaho, but in the Fall of 1905 construction began on the extension to the Park boundary. The road practically follows the valley of Henry Fork of Snake River to Rea Pass, where it crosses the Continental Divide from Idaho into Montana about twelve miles from the terminus.

Snake River Valley is an ideal ranching country readily irrigated and at no distant day will be one of the richest agricultural districts of the West. The Completion of this railroad to the great National playground affords a market for this section that is unequalled.

In the vicinity of Rea Pass, eight miles south of Targhee Pass and Henry Lake, where the railroad leaves the valley for its ascent of the mountains, is a most remarkable cold water spring. The waters of the "Big Spring" form a river fully as large as the stream that flows out of Henry Lake. The wagon road from Snake River valley to the Park will remain through Targhee Pass and the Madison timber reserve to Grayling Inn on the south fork of the Madison.

Graying Inn will be headquarters for tourists until the new hotel on the main Madison is erected.

CHRISTMAS TREE PARK NEAR WESTERN BOUNDARY.

MADISON GRAYLING.

Entering the reservation at Christmas Tree Park, the stage road winds through a beautiful pine forest passing the Riverside Military Station (one of the outposts for the protection of the park) following the Madison River through one of the typical Mountain Canyons until it intersects the belt park road five miles from the Fountain Hotel, Lower Geyser Basin.

At the junction of the Gibbon and Firehole Rivers, the source of the Madison River, is National Park Mountain, historical from the fact that while the Washburn exploring party was camped there on Sept. 19, 1870, it was suggested by a member of the party (Mr. Cornelius Hedges) that this section should be set aside as a National Park, and it was through the efforts of members of the party and others that Congress in 1872 set aside this region as YELLOWSTONE NATIONAL PARK.

FOUNTAIN HOTEL.

The Fountain Hotel.—This elegant and modernly constructed hotel is pleasantly situated on the east side of the valley, commanding an extended view of the surroundings. Its appointments are first class throughout, electric light, steam heat, and the only hotel in the Park having natural hot water baths. It is the first hotel reached by visitors entering the Park from the west. The adjacent streams are stocked with "Loch Leven" and "Eastern Brook" trout, and with the many natural curiosities in this vicinity one can profitably spend several days at the "Fountain."

LOWER GEYSER BASIN.

This is a comparatively wide valley, extending southward from the junction of the east fork of Firehole River with the main stream, and embracing an area of thirty to forty square miles. Over this valley or basin are scattered hot springs in groups, of which Dr. Hayden, in his official survey of the Park Region, has catalogued 693, exclusive of 17 geysers. The central portion of the basin is a nearly level plateau, six or seven miles in width, only partially timbered, and covered with either spring deposit or marsh. The general elevation is about 7,250 feet, while the surrounding slopes, which are, for the most part, heavily timbered, are 400 to 800 feet higher.

Fountain Geyser occupies an eminence, south of the hotel about 2,000 feet. Its eruptions can be seen from the hotel, but not as satisfactorily as when one is standing near the geyser, the great quantities of steam arising from its overflow on the hotel side obstructing the view. The formation or deposit from the waters of this geyser covers an area of several acres, the crater of which is thirty feet in diameter, surrounded by a rim-like edge, to the margin of which the water rises, except upon the south side, where a mound of beaded geyserite has been built up to a height of three or four feet. On the north side of the geyser proper is a considerably larger pool which receives the overflow from the crater. The cushion-like masses of geyserite, which are plainly visible through the transparent blue water, in both the crater and the pool, are very much admired. The eruptions of the Fountain occur at intervals of from two to four hours, and continue with great force from ten to fifteen minutes. During activity the main volume of water does not reach a height beyond fifteen or twenty feet, though jets are frequently thrown fifty or sixty feet. Indications of an eruption are as follows: When both the pool and crater are full of water to the rim it is probable that an eruption will soon take place, as immediately after action the water falls from twelve to eighteen inches below the crater rim, from which point it rises gradually until the climax is reached.

In July, 1899, the Fountain Geyser ceased operations and remained inactive until October, when it resumed its usual displays. In the meantime an immense geyser broke out in the large pool north of the Fountain. Its eruptions were of great force, quite irregular, but gave exhibitions equal to Old Faithful, only continuing, at times, fully an hour.

Clepsydra Spring, some fifty feet west from the Fountain, has recently developed into an active geyser of no small eruptive power, its frequent displays being really quite violent for so small a "spouter" and very pleasing withal.

Mammoth Paint Pots.—Some few hundred feet east of the Fountain, near the road, from which they are separated by a fringe of trees, are situated these won-

MAMMOTH PAINT POTS.

derful paint pots. This remarkable mud caldron has a basin which measures 40x60 feet with a mud rim on three sides, which is from four to five feet in height. In this basin is a mass of fine, whitish substance which is in a state of constant agitation. It resembles some vast boiling pot of paint or bed of mortar with numerous points of ebullition; and the constant boiling has reduced the contents to a throughly mixed mass of silicious clay. There is a continuous bubbling up of mud, producing sounds like a hoarsely whispered "plop-plop," which rises in hemispherical masses, cones, rings and jets. On the north side of the mud basin the rim is low and forms the edge of a flat of pink and red, which is cracked and seamed, and over which are scattered thirty or forty mud cones, generally of a pink and rose color—though a few are gray—averaging from two to three feet in height.

There are two roads from the Fountain Hotel to Excelsior Geyser—one leading west and following south along the Firehole River, the other south in the vicinity of the Great Fountain and skirting along the base of the cliff that terminates opposite the Excelsior.

The Great Fountain is situated about two miles south of the hotel and about one mile east of the main road; the one to the geyser branches off soon after crossing Fountain Creek. The Great Fountain, as described by Mr. David E. Folsom, who witnessed a display October 1, 1869, faithfully portrays its present exhibitions. "The hole through which the water was discharged was ten feet in diameter, and was situated in the center of a large circular shallow basin into which the water fell. There was a stiff breeze blowing at the time, and by going to the windward side and carefully picking our way over convenient stones we were enabled to reach the edge of the hole. At that moment the escaping steam was causing the water to boil up in a fountain five or six feet high. It stopped in an instant, and commenced settling down—twenty, thirty, forty feet—until we concluded that the bottom had fallen out, but the next instant, without any warning, it came rushing up and shot into the air at least eighty feet, causing us to stampede. It continued to spout at intervals of a few moments for some time, but finally subsided." Eruptions of the Great Fountain occur about every ten or twelve hours, the display lasting fully thirty minutes, and it is frequently seen immedi-

GREAT FOUNTAIN GEYSER.

ately after an eruption of the Fountain. In the shallow basins around the crater are seen small particles of geyserite, worn nearly or quite round; these "geyser eggs" are a production of this mammoth spouter only. The indications of an eruption are quite reliable; thirty minutes from the time the crater and basins are filled with water, and the same begins to overflow, the eruption takes place. There are many very interesting and curious sights in the vicinity of the Great Fountain that should be visited. The "White Dome," "Surprise," "Firehole Spring," "Mushroom," and "Buffalo Spring" are the most prominent. The latter was discovered in 1869 by an early exploring party. In describing their trip the writer says: "In one of these springs we saw the whitened skeleton of a mountain buffalo that had probably fallen in accidentally. No king or saint was ever more magnificently entombed that this monarch of the hills in his sepulchre in the wilderness."

Midway Geyser Basin.—Strictly speaking, this section constitutes the upper portion of the Lower Basin, and is about three miles from Fountain group. Being about midway between the extremes of the Upper and Lower Geyser Basins, this locality is given a distinct designation.

Excelsior Geyser.—"Early explorers in this locality discovered, in 1871," says Dr. Peal, "on the west bank of Firehole River, an immense pit of rather irregular outline, 330 feet in length by 200 feet in width at the

EXCELSIOR GEYSER.

widest part. The water is of a deep blue tint, and is intensely agitated all the time, dense clouds of steam constantly ascending from it. It is only when the breeze wafts this aside that the surface of the water, which is fifteen or twenty feet below the level surrounding, can be seen. The walls on three sides are perpendicular, cliff-like, and in places overhang, having been worn away on the other." Cliff Caldron, with every indication of a powerful geyser with long intervals of eruption, was, however, not known to be a geyser until some ten years later. Visited by thousands annually, this section became known as "Hell's Half Acre," a name it retained until 1881, when discovered by Colonel P. W. Norris to be a geyser of great force, and then named by him "Excelsior." Its eruptions in 1881 began in the fall, after the tourist season had closed; Colonel Norris witnessed upwards of thirty eruptions, varying from 75 to 250 feet in height, at intervals of one to four hours. No further eruptions of this geyser are recorded until early in the spring in 1888, when reports became current that Excelsior was again in action. Eruptions of great force continued during the spring and summer in 1888, which resulted in enlarging the crater fully 100 feet. The intervals of eruption during 1888 were at first about every hour and fifteen minutes, increasing towards the latter part of the season to two hours. The only possible indication of an approaching display was the increase in the volume of overflow, there being a steady filling of the crater after periods of activity. Immediately preceding an eruption a violent upheaval occurred, raising the entire volume of water in the crater nearly fifty feet, then instantly one or two and sometimes three terrific explosions would occur, followed closely by the shooting upwards of columns of water, and oftentimes masses of the rocky formation, to a height of 200 to 250 feet. Tons of rock have in this way been hurled into Firehole River, some pieces fully 500 feet from the crater. At each upheaval sufficient water would escape to raise the river several inches. The inactivity, during 1888, of two of the largest geysers in the Upper Basin was attributed to the wonderful activity of Excelsior, which, at each eruption, ejected as much water as all the geysers combined.

Turquoise Spring, situated about 150 feet north of Excelsior, is a silent pool, about 100 feet in diameter, and remarkable for its beautiful blue transparent water. There is a constant overflow from the spring, through a shallow-channel some two feet wide, its sides and bottom being exquisitely colored; when Excelsior was in action the water in this spring sank fully ten feet and did not resume its normal condition for nearly a year. West of the Turquoise Spring, and in itself a marvel, is a small spring of cold water, which, though rather "brackish" to be palatable, is attractive as being the sole cold spring in this region of thermal waters.

Prismatic Lake is probably the very largest and certainly one of the most beautiful springs in the entire Park Regions. It is situated some 500 feet or so west of Excelsior Geyser, its dimensions being 250x400 feet. Over the central pit, or bowl, of this spring the water is of a deep blue color, changing to green towards the margin, while that in the shallower portions of the lake surrounding the central basin has a yellow tint gradually fading into orange. Outside its rim there is a brilliant red deposit, which shades into purples, browns, and grays, all seemingly painted upon a ground of grayish white, which forms the mound, built up of layers of silicious deposit, upon which the spring is situated. This coloring is in vivid bands, which are strikingly marked and distinct. The water flowing cff in every direction, with constant wave-like pulsations over the artistically scalloped and slightly raised rim of the lake, has formed a succession of terraces, each a few inches in height, down the slopes of the mound, particularly upon its southern face. It is impossible to exaggerate the delicacy and richness of the coloring in and about this wonderful phenomenon of nature. The temperature of the water is about 146 degrees Fahrenheit, and the constantly rising clouds of steam sometimes render difficult a good view of the lake surface; but viewed from the proper standpoint (generally with the sun to the back), these same volumes of steam are exceedingly attractive, reflecting the colors of the rainbow or prism, whence the name of the spring, though some attribute it to the variegated tints of its waters. The entire drive from Midway to the Upper Basin, some five miles, is among these natural wonders, but tourists usually proceed to the hotel located at the extreme south end of the Upper Basin before beginning a minute and detailed examination of them.

UPPER GEYSER BASIN.

The basin is triangular in form and embraces an area of about four square miles; it contains twenty-six geysers and upwards of 400 hot springs. Iron Spring Creek bounds it on the west; timbered mountain slopes, extending from southeast to northwest, form the hypotenuse of the triangle, and a wavy line of dark forest conifers, its southern base. The main Firehole River drains it, centrally; its shelving banks are thickly pitted with steaming hot springs and studded with mounds and cones of geyserite. Here, grouped within the narrow space of, perhaps, a square mile, are the grandest and mightiest geysers known to man, and silent pools of scalding, meteoric water that for beauty of formation and delicacy of coloring are simply marvels. The surface of the basin consists, for the most part, of a succession of gentle undulations, each crowned with a geyser-cone or hot-spring vent and covered with layers of silicious sinter and crumbling carbonaceous deposits, that give it a grayish-white, sepulchral hue. Clouds of vapor hang shroud-like above it; the earth trembles and is filled with strange rumblings, the air is heavy with sulphurous fumes, and vegetable life is extinct. In a paper read before the Cardiff (Wales) Naturalists' Society, Mr. Charles T. Whitmell said: "Nowhere else, I believe, can be seen, on so grand a scale, such clear evidence of dying volcanic action. We seem to witness the death throes of some great American Enceladus. Could Dante have seen this region he might have added another terror to his Inferno."

And, continuing, the same writer quotes Lord Dunraven, as saying that a view of Firehole Valley gave him the impression that some modern cities had been overwhelmed, and had so lately sunk amid flames into the bowels of the earth that the smoke of their ruins was still ascending through heaps of smouldering ashes.

GEYSER TIME TABLE.

A RECORD OF THE ERUPTIONS OF THE GEYSERS IN YELLOWSTONE PARK, DURING THE SEASON OF 1905.

UPPER BASIN.	HEIGHT FT.	INTERVALS OF ERUPTION.	DURATION.
Old Faithful	150	65 to 75 minutes.	4 min.
Chinaman............	40	Irregular.	2 "
Cascade............ ..	30	Irregular.	1 "
Bee Hive	200	12 hours to 4 days.	8 "
Giantess........... ..	150	16 to 25 days.	12 hours.
Lion...........	60	3 to 20 hours.	8 min.
Lioness	80	1 to 4 days.	10 "
Cub.	12	Frequently.	20 "
Surprise.............	100	Irregular.	2 '
Sawmill.	35	Frequently.	30 '
Spasmodic	40	Irregular.	20 '
Grand...............	200	12 to 24 hours.	30 '
Turban	40	Frequently.	20 "
Economic....	20	3 to 20 minutes.	10 sec.
Oblong..............	35	6 hours to 2 days.	5 min.
Giant	250	6 to 12 days.	90 "
Grotto	30	3 to 4 hours.	30 "
Riverside	100	8 hours.	15 '
Mortar	30	2 hours.	5 '
Fan	60	4 to 6 hours.	10 '
Artemisia	50	12 to 24 hours.	10 '
Jewel	30	5 minutes.	1 '
Splendid.............	200	3 hours to 2 days.	10 "
Comet.	60	Irregular.	5 '
Castle	75	24 to 30 hours.	30 '
Cliff	100	4 to 8 hours.	8 '
Lone Star............	75	40 minutes to 2 hours	10 "
Daisy	60	2 hours.	7 '
LOWER BASIN.			
Fountain.......	50	3 to 5 hours.	20 min.
Great Fountain	100	8 to 12 hours.	30 "
NORRIS BASIN.			
Constant.............	40	1 to 2 minutes.	10 sec.
New Crater	20	5 to 15 minutes.	2 min.
Monarch	75	3 to 4 days.	30 "

OLD FAITHFUL INN FROM THE BEE HIVE.

Old Faithful Inn, the most extensive log structure yet devised by man, with every convenience and luxury of the modern hotel, is the latest triumph in utilizing primitive material in constructing so unique a building. The rough blocks of stone which form its foundation appear as natural as when found at the base of the cliffs of the surrounding mountains.

The interior is surprising for the grandeur of vast forests conquered. Massive logs tapering on each ascending balcony appear as giant trees. The staircase leading to the lookout has split logs for steps. Windows of diamond shaped panes and dainty French curtains are exquisitely beautiful against the setting of rough logs. Elbows of natural branches gathered from the neighboring forests form the braces for the numerous gables and frame the many balconies and stairways surrounding the office while timbers braced this way and that, support the high roof. The huge doors of the entrance and dining room are noticeable to all. Their hinges and quaint iron locks, together with the immense clock, were hand forged from bar iron. The old time fire place is a welcoming sight and its chimney, sixteen feet square at its base and made of large lava blocks, towers high through the roof four stories above. The chimney contains four large and four small fire places and fastened against it is the immense clock keeping mountain time. In another chimney in the dining room is constructed the old time spit and oven. In contrast to the rough logs there appear electric candles, hardwood floors, Mission furniture and gay rugs and curtains, all of which give a warmth and richness to the building. The bell on the roof announces interesting and extraordinary events and tolls a quarter of an hour before the opening of the dining room. The center of the building rises eight stories high, surmounted by the lookout that gives a panoram ic view of the geyser basin. From half a dozen golden topped flag staffs float the emblems of various nations. At night, by the aid of a powerful search light, one discerns geysers in action, bear feeding at the edge of the timber, while the illumination of Old Faithful Geyser in action is a sight never to be forgotten. The Old Faithful Inn was built at a cost of two hundred thousand dollars and was open to the public for the season of 1904.

OLD FAITHFUL GEYSER.

The more detailed description given on page 55 of the chief geysers and springs will serve to acquaint the readers with the peculiar characteristics of each.

Old Faithful.—Less than 1,000 feet east, and in plain sight from the hotel, is located this reliable friend of the tourist. Every sixty-three minutes (with rarely a variation of five minutes) day and night, summer and winter, this wonderful freak of nature gives its exhibition. The position and direction of the sun and wind vary the appearance of this geyser, which is one of the most popular in the Park, because of the remarkable regularity with which its eruptions occur, and the excellent opportunities afforded for observation. Eruptions by moonlight, at sunrise or sunset, in a storm or with clear weather, with their varied effects, command the attention of the visitor, regardless.

Its eruptions begin with a few spasmodic spurts, during which considerable water is thrown out; these are followed in from five to eight minutes by a column of hot water two feet in diameter, which is projected upwards to a height of 125 to 150 feet, when it remains apparently stationary for about three minutes. The crater is an oblong opening two by six feet on the inside and four by eight on the outside, is situated on a mound of geyserite, measuring at the base 145x215 feet, at the top 20x54 feet, the whole rising about twelve feet above the level surrounding. This mound is composed of layers of deposit in a succession of distinctly marked terraces which are full of shallow, basin-like pools, the water in which is clear as crystal, and their edges or rims exquisitely beaded and fretted, their bottoms showing delicate tints of rose, white, saffron, orange, brown and gray. The north end of the crater has large globular masses of beaded, pearly deposit, and its throat is of a dark yellow or rusty color.

CASTLE AND BEE HIVE GEYSER.

Bee Hive.—Crossing the footbridge leading to the east side of Firehole River, this geyser is found about 100 feet from the river bank. Its name was suggested by the peculiar shape of its cone, which is about four feet in height, three feet in diameter at the top, by seven at the base, and nearly circular. Its nozzle-like opening, or crater, is about eighteen inches across at the apex, narrowing gradually till the base of the cone is reached. The Bee Hive's eruptions are irregular, generally occurring about three times daily. It has, however, been known to have periods of activity of not more than three hours apart, and, per contra, to remain inactive for several weeks. There are, usually, several eruptions, about three hours apart, immediately following the Giantess. There is no terrace-shaped deposit surrounding this geyser, as is the case with most of the others, and it is the only one close to which persons can approach with perfect safety while in action. So hot is the water ejected that it, for the most part, evaporates while in the air. The height attained varies from 170 to 220 feet.

A miniature geyser, or indicator, a few feet from its base, is, generally speaking, a faithful forerunner of activity of the Bee Hive, by shooting up jets or spurts of water, which are followed in about fifteen minutes by a column of steam and water from the main crater, hurled upwards with great force and in a steady stream.

Giantess.—Some 300 feet east from the Bee Hive, upon the highest point of elevation in this portion of the Upper Basin, is located the Giantess, considered by many a geyser of unusual importance, whose eruptions, occurring at intervals of several days, none should fail to see. However, it is necessary that, in order to witness and fully appreciate its entire display, one remain in its vicinity at least twelve hours. Its crater, bowl-shaped, and some sixty feet in depth, is 24x30 feet in size at the surface, and is wholly devoid of the highly colored ornamentation and cone so characteristic of other geysers in this region. As the crater rapidly fills with water after an eruption, it resembles, to most visitors, a large slightly agitated pool of sapphire-tinted water, with no outward indication of being the powerful geyser which it really is. At the beginning of an eruption the entire contents of the crater are instantly forced out, flooding the whole region round about. Relieved from this immense pressure of water, the geyser at once begins to eject forked-like columns of water and steam into the air, throwing them to a height of from 60 to 100 feet above the surface. These displays continue at short intervals throughout a period of about twelve hours, or until the water in the geyser tube is entirely exhausted, when an interesting "steam period" takes place, lasting nearly an hour, and producing a roaring sound audible in all parts of the basin, and when occurring at night often awakens the guests at the hotel. In the earlier stages of eruption, during which the emptying of the crater takes place, shocks, similar to those produced by earthquakes, are distinctly felt throughout the basin, while the disappearance of adjacent springs, and the generally succeeding activity of the Bee Hive, give rise to the theory of subterranean connections between geysers and springs upon this "bench."

The Sponge.—A short distance to the north of the Giantess is a curious little geyser called the Sponge, whose slightly raised, circular cone, strongly resembling a huge sponge in the character and color of its formation, attracts the eye of the passerby.

Lion, Lioness and Cubs.—The Lioness and Cubs occupy a conspicuous mound of geyserite to the west of the Lion, which has an irregular flat-topped cone about four feet in height, and is separated from the rest of the group by a slight depression. Eruptions of the Lion occur daily, those of the Lioness not as regularly, the Cubs play more frequently, and it often happens that the Lioness and Cubs play together, though it rarely occurs that the Lion and Lioness are seen in eruption at one and the same time. The former is the most powerful of the group, and throws a column of water fifty or sixty feet high, frequently continuing in action ten or fifteen minutes.

Sawmill Geyser.—Leaving the group just described, the pathway leads down the basin, passing through a point of timber close by the river, emerging from which one is confronted by a number of active springs and small geysers (situated upon a "bench" similar to the one just left), among which are the Tardy, Bulger, Spasmodic and Sawmill, the last named being the chief attraction, as well as the largest in the group. This locality may be also reached from the west side of the river by means of a footbridge near the Castle.

The basin of the Sawmill is shallow, and about forty feet in diameter, inclosing another basin of about half its size, in the center of which is located its funnel-shaped crater, some seven feet across the top, and sloping to a small orifice. Its eruptions are very frequent, usually five or six a day, each continuing in operation fully an hour. The peculiar noise accompanying activity (suggesting its name) coupled with its spiteful vigor, renders this little spouter quite attractive.

The Grand and Turban.—At the base of a rocky bluff, some 500 feet northeast from and nearly opposite the Castle, are situated the above important geysers. An observer is naturally led to suppose that the irregular, pit-shaped crater, noticeably prominent, is that from which the Grand plays; such is not the case, however, as this crater is merely a water basin or reservoir, undoubtedly having connection with the Turban, but entirely unconnected with the Grand, which plays from an opening situated a few feet to the south, surrounded by cushion-like masses of geyserite formation. Eruptions of the Grand are somewhat irregular, usually occurring at intervals of from twelve to twenty-four hours. The outbursts of the Grand are among the finest to be witnessed in the Park, having a series of ten or twelve distinct eruptions, lasting from thirty to forty minutes, and throwing at each discharge forked columns of water to a height of 200 feet, allowing ample time for visitors who may chance to be in any other part of the basin, to reach its vicinity in time to witness some, at least, of its several magnificent displays. The Turban plays, mainly, from a fissure-like opening in its formation, immediately north of its main crater, which, meantime, is greatly agitated, often overflowing, and discharging quite large quantities of its hot flood into the crater of the Grand, just below it on the south. The frequency of the Turban's eruptions occasionally presents the unusually fine spectacle of both geysers, Grand and Turban, in action at the same time. On the way from this point to the Giant and Grotto (reached by a footbridge across the Firehole near the

Oblong), Beauty Spring and Economic Geyser are passed, the former one of the large silent pools of the Upper Basin, remarkable for the vivid coloring and exquisite beauty of its highly ornamented margin, and its limpid blue waters.

Economic Geyser, named from the curious fact that there is no overflow nor waste whatever from it, as its waters, though frequently thrown fifteen or twenty feet in the air, fall again into the crater and disappear. It is a frequent surprise to visitors who happen along when it is quiet, to see an eruption so sudden and spiteful, occurring without warning, at intervals varying from three to twenty minutes.

EARLY MORNING, UPPER GEYSER BASIN.

GIANT GEYSER.

The Giant.—It is nearly a mile from the hotel to this monarch of geysers, situated, like many others, in close proximity to Firehole River. Its cone, about ten feet in height, though some 200 feet from the roadway, is conspicuous. A few feet to the north is an irregular mound, from which considerable steam escapes from sundry small holes, undoubtedly connected in some manner with the geyser and acting as its escape valves. The platform of deposit upon which the cone stands is about seventy-five feet in diameter. The cone is broken on the west side from the apex nearly to its base, affording a good view of the interior of the crater, which is almost constantly in a state of turbulent boiling and splashing. In 1881 the break in the cone was not nearly so large, not more than half its present proportions, the enlargement being without a doubt the result of an unusually violent eruption.

The Giant usually "plays" about once a week, for a period of one and one-half to two hours. An immense column of water is lifted 250 feet into the air at its initial outburst, the height of the same gradually decreasing until the close of the display, which is preceded by a rumbling sound not unlike a distant train of moving cars. Those fortunate enough to have been an eye-witness to one of its gigantic displays have enjoyed a treat most rare.

The Oblong.—Quite near the river bank, a short distance south from the Giant, is situated the Oblong. Its crater is about 30x50 feet in size—hence its name—the interior of which, immediately following an eruption, is exposed to a depth of several feet, and, lined as it is with large globular masses of formation, affords the finest view of interior geyser structure to be seen in the entire Park region. Two large openings can be clearly seen in the bottom of the crater, and when the water is not agitated the eye readily penetrates them to unknown depths. Eruptions of the Oblong vary from six hours to two days, lasting only a few minutes, during which the contents of the crater are raised bodily some twenty feet.

The Grotto.—By far the most curious geyser cone of all is that of the Grotto, situated close by the roadside, some 500 feet northwest of the Giant. The various cave-like openings in its peculiarly shaped cone give rise to its name. Its eruptions take place about four times daily, each display lasting fully thirty minutes, though, owing to its singular construction, its waters are not thrown to any considerable height (scarcely twelve feet). During eruptive periods, however, immense volumes of steam escape with great force. Separated from the main cone some twenty feet is a smaller crater which acts with the main geyser during eruptions.

Riverside Geyser.—A short distance above the wagon bridge across Firehole River is seen the Riverside Geyser, whose cone is close to the water's edge; it consists of two chimney-like craters, the larger being at the same time the higher. The geyser "plays" from the lower opening only, though visitors are apt to arrive at a reverse conclusion when viewing the locality between periods of eruption. An overflow of water from the lower crater is a certain indication of approaching activity, beginning about thirty minutes previous to eruptions and continuing until the outburst, which takes place about every eight hours, throwing an arching column to a height of eighty or ninety feet, the entire contents of the discharge falling into the river.

The Fan and Mortar.—On the east bank of the Firehole, about 300 feet below Riverside Bridge, are located the Fan and Mortar Geysers. The former has an eruption every eight hours, frequently following the Riverside, its ejected waters spreading out in fan-shaped jets, from the fact of its having two crater orifices which throw out diverging streams. The pink geyserite forming its crater is quite unlike that of any other geyser. The Mortar, a few feet to the south, "plays" more frequently than the Fan, and, when viewed from the bridge above alluded to, resembles in its eruptions that particular piece of ordnance from which it derives its name.

The Splendid.—One of the most remarkable geysers in the Upper Basin is the Splendid. It is located fully 1,000 feet west of the Giant and a short distance south from a prominent mound of geyserite, called, from its color, the White Pyramid. The entire absence of anything like cone structure, and the numerous crater-like openings in its vicinity, puzzle one at first to locate the geyser proper; however, the extreme western opening, nearest the knoll and timber, is that from which the Splendid plays. A peculiar feature of this geyser is, it will have several eruptions about three hours apart and then remain inactive for a couple of days. When in action it throws a powerful stream fully 150 feet in height, increasing in force very perceptibly during the earlier stages of its eruptions, and not reaching the climax for several minutes, apparently maintaining its greatest vigor from five to eight minutes. Quite unlike other geysers, the Splendid throws its stream at a sharp angle instead of vertically, which fact, when it was first discovered, caused it to be called the Comet; this designation, however, soon gave way to its present more appropriate appellation. During the afternoon eruptions, if the sun be visible, highly colored rainbows add to the rare beauty of the displays, and when seen, as it occasionally is, in conjunction with eruptive activity of other small geysers of the Splendid

CRATER OF OBLONG GEYSER.

group, produces a truly marvelous effect. No perceptible change in the appearance of its crater follows or precedes periods of activity, and even during its quiet days the same violent boiling of its waters is always the subject of noticeable comment.

The Punch Bowl.—The wagon road leading westward from the Splendid toward Black Sand and Sunlight Basins passes the Punch Bowl, by far the handsomest spring of its peculiar class to be found in the geyser region, if not in the world. Situated on the summit of a small mound of silicious deposit, some five feet above the general level, it is about ten feet in diameter, with a glittering rim of brilliantly colored formation eighteen inches in height. The constant boiling of its contents, though only a small part of its surface is agitated, as the bubbles of escaping steam and gas arise, produces a wave-like undulation over the entire spring and gives it a steady and not inconsiderable overflow. A small cave-like opening on the east side of the mound or cone is very handsome and much admired, having the appearance of being lined with satin of the rarest beauty and texture. Early visitors to the Park during the seasons of 1873 and 1875 speak of this spring as being an active geyser, and during 1888 similar reports gained currency. Nothing, however, is certainly known as to the correctness of these reports, though they are highly probable.

Black Sand Basin and Specimen Lake.—Dr. Peale's description of Black Sand Spring is interestingly comprehensive, and is as follows: "This is one of the most beautiful springs in the Upper Basin. It has a delicate rim, with toadstool-like masses around it. The basin slopes rather gently toward a central aperture that, to the eye, appears to have no bottom. The water in the spring has a delicate turquoise tint, and as the breeze sweeps across its surface, dispelling the steam, the effeet of the ripple of the water is very beautiful. The sloping sides are covered with a light brown crust; sometimes it is rather a cream color. The funnel is about forty feet in diameter, while the entire space covered by the spring is about 55x60 feet, outside the rim of which is a border of pitch-stone (obsidian) sand or gravel sloping twenty-five feet. From its west side flows a considerable stream, forming a most beautiful channel, in which the coloring presents a remarkable variety of shades; the extremely delicate pinks are mingled with equally delicate tints of saffron and yellow, and here and there shades of green." The overflow from this spring spreads out over a large area, called Specimen Lake, which deserves more than passing notice. Absorption of the surrounding silica has destroyed many of the trees in the vicinity, the dry, lifeless trunks adding to the attractiveness of the place, geologically speaking, by affording the appearance of petrifications.

Leaving Black Sand Basin, the next attraction is Sunlight Basin and Emerald Pool, reached by a footbridge over Iron Spring Creek.

Sunlight Basin, composed of several large, silent pools, whose coloring far surpasses anything seen in the Park. The most fascinating spring in this collection is Emerald Pool, the most southern one in the group. It is frequently spoken of by visitors as the most beautiful object in the geyser basins. Cliff Spring, a few feet south of the footbridge, on the very edge of Iron Spring Creek, with its wall of formation built at right angles around the spring, is the first to attract the visitor.

CLIFF SPRING.

A road from Sunlight Basin leads to the hotel, passing some ancient geyser cones and the "Three Sisters," a trinity of springs quite attractive. Or one can retrace the road just followed, intersecting the main driveway near the Splendid Geyser.

Castle Geyser.—The Castle is at once recognized, as it occupies a very prominent position close by a point of timber on the main road midway between the Splendid Geyser and Old Faithful, visible from nearly all points of the basin. The great amount of deposit, perhaps 100 feet in diameter, at its base, and the possession of the largest cone in the whole region, while giving it an air of conspicuousness, at the same time indicate that it is one of the oldest active geysers in the Park.

The broken condition of its cone on the east side renders possible an easy ascent to its summit, which is about twenty feet across. The orifice of the geyser tube in the top of the cone is about three feet in diameter, quite round, and is lined with a formation of bright orange color. Eruptions of the Castle occur at intervals of about thirty hours, preceded by the occasional throwing out of jets of water to the height of fifteen or twenty feet, perhaps. These premonitory symptoms of eruption generally continue five or six hours, when more violent demonstrations, during which columns of water are shot upward to a height of fully seventy-five feet, ensue, and, continuing for half an hour or so, are followed by a "steam period" similar to that of the Giantess. Several times each season it has eruptions of an unusual character, in which its columns of water are thrown to twice their usual height and its subsequent "steam periods" are proportionately forcible. A violent boiling spring is situated near the base of its cone, on the north side, and used to be a favorite resort of the "camper-out" in earlier days, as excellent coffee can be prepared in this spring in fifteen minutes, and other edibles, requiring the action of boiling water to prepare them for the table, are well and thoroughly cooked in a correspondingly short time. The large, crested spring, 100 feet north from the Castle, is usually very handsome. It generally is filled to overflowing, and the bottom and edges of the channel leading out of its north side are very highly and beautifully colored. This spring is some twenty feet in diameter, and is commonly known as Castle Well.

Morning Glory Spring.—A few hundred yards below Riverside Bridge a road branches off the main thoroughfare, passing around the Morning Glory Spring, which can be seen from a stage or carriage, a slightly elevated point of observation affording a better view. It is a silent pool some twenty feet in diameter, overflowing slightly at the west. The peculiar shape of its funnel-like crater, whose walls are delicately colored, together with the beautiful transparency of its waters, suggests its very appropriate name. Half a mile below is

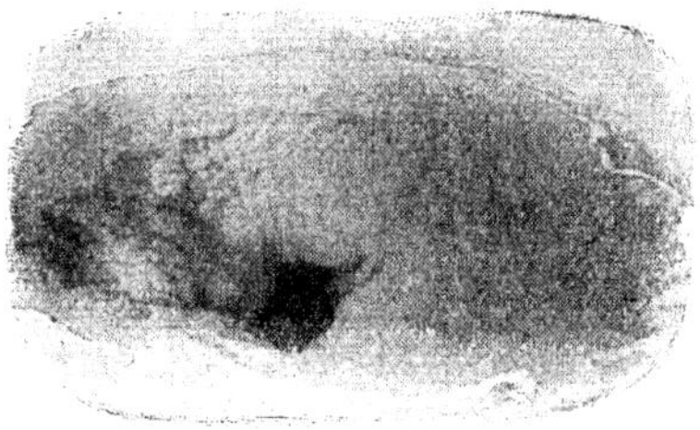

MORNING GLORY SPRING.

Artemisia Spring, situated between the road and the river, quite near the former, which is elevated some twenty feet above the spring. Stepping to the edge of the bank, an excellent view of the crater is obtained, the crystal clearness of its waters allowing a distinct view into its apparently bottomless depths. The spring is sixty feet in diameter and generally very little agitated, merely overflowing. The surrounding formation, quite unlike that of any other spring or geyser, is as hard as flint, and of a peculiar olive-green color. Although for the most part very quiescent, this spring has occasional pulsations in the nature of eruptions, at which times large quantities of water are forced out, fairly flooding the formation between it and the river. These eruptions occur at intervals of twelve to twenty-four hours. The bank of the Firehole, some thirty feet high at this point, is the most highly colored section of the river to be found in the Upper Basin. The best view is obtained from the bridle-path on the opposite side of the river. This trail leads south from the Splendid, crossing the Firehole just above its confluence with Iron Spring Creek, near which it joins the main road.

Biscuit Basin is on the west side of Firehole River and on the north side of Iron Spring Creek, being about one mile below Riverside Bridge. The footbridge over Firehole River nearby allows one to easily visit this locality. The principal attraction in Biscuit Basin is Sapphire Pool, whose highly ornamented margin suggested the basin's rather odd name. Hundreds of small symmetrical, biscuit-like knobs of olive-green formation surround the spring, which is of the variety known as pulsating or breathing springs (geysers in fact). The constant ebb and flow of its waters have produced this peculiar formation, from one to another of which one must pick his way in order to get a good view of the pool itself. A few feet to the west is

Jewel Geyser, whose eruptions occur with the remarkable frequency of from three to five minutes, throwing its jets of water to a height of twenty-five or thirty feet. Scarce 500 feet further west are the Black Pearl and Silver Globe. The former has a beautiful basin, studded thickly with black pearls, each about one-quarter of an inch in size. A curious feature of this little "spouter" is the fact that its formation surrounds

BISCUIT BASIN, SAPPHIRE POOL.

the roots and stump of a tree, completely incrusting the same with its rich, black ornamentations.

The Silver Globe derives its name from the constant rising to its surface of large, silvery globules or bubbles of gas or steam, which, of course, immediately disappear on reaching the air.

UPPER GEYSER BASIN TO YELLOWSTONE LAKE.

The third day of the regular tour of the Park is consumed in traveling from the Geyser Basin to Yellowstone Lake. The route is over the summit of the continental divide, near Shoshone Lake, the head waters of Lewis Fork of Snake River, a branch of the Columbia that empties into the Pacific Ocean; and in a few miles returns to the Atlantic Slope at Yellowstone Lake, whose waters reach the ocean through the Yellowstone, Missouri and Mississippi Rivers.

Leaving the geysers, the road follows up the Madison River (being the same stream known as the Firehole River during its meandering of the Geyser Basins), crosses the same and climbs a gentle ascent to

Keppler Cascades, about a mile and one-quarter distant, whose waters leap from shelf to shelf of a rocky chasm in a series of enchanting falls, aggregating 100 to 150 feet in height, and whose charms are enhanced by the dark background of forest on either hand. The roadway continues up the Madison about two miles to the third crossing, when it leaves the river, following

the course of Spring Creek nearly to the summit of the divide.

At the third and last crossing of the Madison a side trip can be made to Lone Star Geyser, about half a mile south of the bridge on the west bank of the river.

Lone Star Geyser.—The cone of this geyser is about ten feet in height and twelve in diameter at its base. Its summit is fully six feet across. The crater consists of one large central opening surrounded by numerous small orifices, from all of which water is thrown during an eruption. The Lone Star plays at intervals of thirty minutes to two hours, its boiling contents being thrown in a fine spray, mingled with steam, to a height of sixty feet. The chief beauty of this geyser lies in its cone, which is striped vertically with bands of white, lavender and brown, intermixed with varying shades of yellow, and is completely covered with an almost endless variety of elegant pearl-like beads. In leaving the Madison the road deflects to the eastward and climbs, with gradual ascent, to the continental water-shed. The first two miles it traverses the winding, rocky canyon of Spring Creek, and finally reaches the summit at "Craig Pass," a narrow, rocky gorge hemmed in by precipitous cliffs. Norris Pass, through which the trail leads to Shoshone Lake, is less than a quarter of a mile to the south.

Shoshone Point, about half way between the Upper Basin and Thumb Bay (Yellowstone Lake), affords one of the most commanding views of the ride. It overlooks Shoshone Lake and its beautiful valley surrounded by heavily timbered slopes. Shoshone Lake has an area of about a dozen square miles, with an irregular shore line. Shoshone Geyser Basin, situated on the west shore of the lake, has several large geysers and numerous interesting springs; it is reached by trail from the Lone Star Geyser.

TETON MOUNTAINS.

On a clear day from Shoshone Point may be seen the three snow-capped "Sentinels" of the Teton Mountains fifty miles distant, that form a portion of the boundary between the states of Wyoming and Idaho, their dizzy heights, full 14,000 feet, overtopping all other peaks of the Rockies.

From Shoshone Point the drive is less attractive; however, it again crosses the continental divide at a "pass" so level that it is difficult to know when the summit is really reached.

Lake View—A mile from the lunch station at Yellowstone Lake, one catches the first glimpse of this beautiful sheet of water. From this point Mr. David E. Folsom, of the Folsom and Cook exploring party in 1869, says: "As we were about departing on our homeward trip we ascended the summit of a neighboring hill and took a final look at Yellowstone Lake. Nestled among the forest-crowned hills which bounded our vision lay this inland sea, its crystal waves dancing and sparkling in the sunlight as if laughing with joy for their wild freedom. It is a scene of transcendent beauty which has been viewed by but few white men, and we felt glad to have looked upon it before its primeval solitude should be broken by the crowds of pleasure seekers which at no distant day will throng its shores."

"Thumb" Bay Lunch Station is about midway between the Upper Basin and the Lake Hotel at the "outlet." Tourists have an opportunity of taking a steamer from here to the hotel if they desire. The drive of sixteen miles is varied, following the lake shore for three or four miles, through a timbered plateau that affords numerous glimpses of this magnificent mountain lake, passing the interesting Natural Bridge, and along Bridge Bay to the hotel.

A wagon road has been constructed south from the Yellowstone Lake, passing Lewis Lake, and continuing down the valley of Snake River to the southern boundary of the Park.

HOT SPRING CONE.

There are no less than seven hot-spring areas surrounding Yellowstone Lake; those of the west arm, or Thumb Bay, are by far the most interesting. They comprise over sixty springs and paint pots and several geyser cones, one of which rises above the lake surface just a few feet from shore, standing upon which one may catch trout, and, dropping them into the hot water in the crater of the cone, cook them without removal from the hook. Some of these springs have considerable overflow, caused by what seems to be a forcing up of their contents, which rise and fall alternately like the bosom of a sleeping giant; these are called breathing or pulsating springs, in contradistinction to those whose waters maintain the same quiet level. The waters of nearly all these springs possess the same delicate blue tints noticeable in other portions of the Park. The lake-shore at this point consists of sloping, terrace-like layers of silicious deposit, which extend some distance back from the water's edge and even out into the lake. Most of the springs are scattered over this formation, back a little way from the lake, though several are close to the water's edge, and a few, even, may be seen beneath the lake surface, occasional points of bubbling betraying their presence. Some 400 or 500 feet back from the lake, and nearly opposite the "fishing cone," is a paint pot basin similar to that near the Fountain Geyser, in the Lower Basin. This basin is about fifty feet in diameter, and is a seething mass of beautifully colored and finely granulated clay, the prevailing tints being pink and red in varying shades, though creamy white and pale blue colors are noticed. Around the edges of the basin are a dozen or more hollow mud cones, two to three feet in height, from which discharges of mud occasionally occur. By many this basin is considered the most attractive of all the paint pots of the Park.

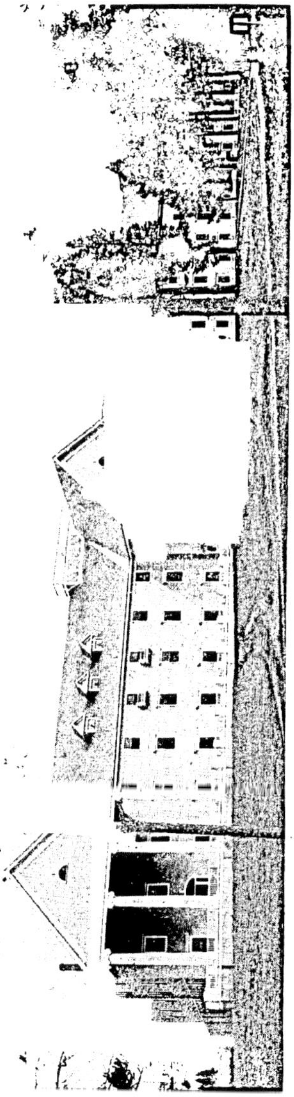

LAKE HOTEL.

Hotel at the Outlet.—This spacious and elegantly appointed hotel tends greatly toward making Yellowstone Lake the resort par excellence of the Park. Here everything is so arranged that guests can spend the entire season, if they so desire, making short, easy trips of sight-seeing or explorations to all points of the great reserve. The falls and canyon are distant but seventeen miles, a well constructed road leading thither; the great geyser basins are scattered along a stage route, whose extreme length is not above forty miles from this hotel, while to the eastward some fifty miles lies the Hoodoo Region or Goblin Land, a weirdly wild region, as yet visited by only a few sportsmen and ambitious explorers, but which time and the construction of roads will render accessible to all.

To visit any or all of the points circumjacent to this grand mountain lake, vehicles of all kinds, saddle and pack animals, guides, rowboats and steamers, are ever at command, and as for trout fishing, he who has never cast a "fly" into the blue depths of this vast natural "trout preserve" and its large river outlet knows little or nothing of its delights. Fifteen by twenty miles in size, of irregular outline, somewhat resembling the human hand, and embracing an area of about 150 square miles, this is the largest body of water in North America at so great an altitude—7,788 feet above sea level.

Several islands dot its surface, the largest being Stevenson, near the south end, and Franks, midway of the lake, and its very considerable depth (from five to fifty fathoms) renders navigation practicable and safe. The Yellowstone River is at once its principal affluent and sole outlet, its upper portion draining a considerable area tributary to the lake on the southeast, and the vast body of water thus accumulated in this natural mountain reservoir serves not only to furnish a never-failing supply for one of the grandest of the Missouri's tributaries, but supplies the means of successful irrigation of the entire lower Yellowstone valley.

THE NATURAL BRIDGE.

The Natural Bridge is four miles southwest from the Lake Hotel, being about one mile

YELLOWSTONE LAKE AND SLEEPING GIANT.

back from the lake. It spans a small creek, a branch of Bridge Creek. The new road from Thumb Bay to Bridge Bay passes Natural Bridge. This interesting arch of stone is some forty feet high and six feet wide, its abutments being some thirty feet apart; when viewed from the lower side it presents the more symmetrical appearance.

The Sleeping Giant.—In the mountain range on the east side of the lake can be seen the "Sleeping Giant." It is formed of the peaks of Saddle Mountain in connection with a mountain range several miles this side.

Fishing Grounds.—In the river at the lake outlet are the fishing grounds about a mile from the hotel, while at many places between the lake and canyon excellent fishing is had from shore. During the trout season (July to September), no better fishing can be found. They average about one and one-half pounds each, and are of the salmo myhiss variety—a catch of 100, three or four hours before sundown, is not infrequent.

Yellowstone **Lake to Sylvan Pass.**—The construction of the road from the outlet of Yellowstone Lake to the eastern boundary of the Park has opened up a delightful portion of Park scenery. A substantial bridge spans the Yellowstone River about two miles from the Lake Hotel. The road passes through the valley of Pelican Creek, along the south shore of Turbid Lake, and gradually ascends the mountains along the foot hills of Avalanche Peak to Sylvan Pass, on the headwaters of Clear Creek. Before the pass is reached one of the finest panorama views of the Park is to be had. Yellowstone Lake occupies the foreground, and the distant snow-clad peaks of the Tetons, Quadrants, Electric Peak, Mount Washburn, and the entire Rocky Mountain range within the Park, outline the horizon. It is about eighteen miles from the Lake Hotel to Sylvan Pass, and a side trip to this rugged portion of the Park is now possible. Excellent camping grounds will be found between the pass and Sylvan Lake, and, until hotel accommodations are provided, it will be necessary to carry camp equipment, as it will require nearly two days to visit all the interesting points of this section.

CODY, WYOMING.

Eastern Entrance to the Park.—The completion of the road through the Eastern Entrance to Cody, Wyoming, a terminus of the "Burlington" system, opens up the most scenic portion of Wyoming to the public. Hotels have been constructed between Cody and the Park for the accommodation of the traveler. Camping outfits and transportation can be secured at Cody for the Park tour which will undoubtedly be largely patronized by lovers of scenery, camp-life and fishing.

Yellowstone **Lake** to **Falls and** Canyon.—The road from the lake to the Grand Canyon follows the valley of the Yellowstone the entire distance, seventeen miles, and most of the way quite near the river. It passes Mud Geysers, Sulphur Mountain, across Hayden Valley, and within four miles of the "National Park Game Enclosure," in the upper valley of Trout and Alum Creeks, where it is intended to protect specimens of all the large animals found in this region—buffalo, elk, deer and mountain sheep.

Mud Geysers are about five miles from the Lake Hotel, and consist of several large craters filled with blue, pasty mud, one and all emitting odors far from agreeable. The attractive feature of this locality is the Mud Volcano, situated a few rods to the west of the road, at the base of the cliff, whose funnel-shaped crater is thirty feet in depth, formed by mud ejected from below through a cave-like opening, out of which a sickening, lead-colored mass of mud, of the consistency of soft mortar, is constantly belched, accompanied by dull, muffled sounds, and in manner at once repulsive and fascinating.

During the summer of 1898 visitors witnessed the most violent eruptions of the Mud Volcano ever recorded. The mud-plastered trees in the vicinity are still in evidence of these terrific explosions.

Hayden Valley extends from Mud Geysers to Alum Creek along the Yellowstone and west from the river

HAYDEN VALLEY.

to Mary Mountain. It is the largest valley in the reservation, and especially adapted as a range for game, being well watered by Trout and Alum Creeks, and protected on the north and west by a heavily timbered range.

Sulphur Mountain consists of a group of isolated hills or "buttes," each about 150 feet high, of which a splendid view is obtained from the roadway that skirts along their western base. Large blocks of detached rock are scattered about, in all of which a great percentage of pure sulphur is noticeable. The fumes arising from the various vents are exceedingly disagreeable. The chief attraction is a large boiling spring at the base of the mountain on the west side. On the opposite side of the road are numerous mud caldrons and springs, the contents of which are varied, some being of thick mud, others of leaden-hued water, all incessantly agitated and throwing upon their surrounding edges a finely mixed muddy deposit. The road from Sulphur Mountain to the Canyon Hotel soon joins the main road along the river, passes over a rolling country, and skirts the banks of the Yellowstone until nearly to the Upper Falls.

Grand Canyon Bridge.—The construction of this magnificent driveway over Yellowstone River, at the head of the rapids above the Upper Falls, affords visitors an opportunity of now viewing the Grand Canyon from the opposite side; also of descending into the Canyon to the foot of the Great Falls, by the only possible trail. The chief attraction is Artists' Point, from which an unobstructed view of the Falls and Canyon is obtained.

GRAND CANYON BRIDGE.

UPPER FALLS OF THE YELLOWSTONE.

About a quarter of a mile above the Upper Falls the current becomes very rapid, tumbling over a succession of cascades and swirling around masses of rock left surrounded in mid-stream. Just before reaching the brink of the falls, the river makes an abrupt turn to the eastward, from which fact an unobstructed view of the falls cannot be obtained from the hotel side. Above the falls a jutting point of rock affords an excellent view of the rapids and the foaming waters rushing on over the precipice itself. The Upper Falls have a perpendicular drop of 140 feet, and the water, striking the shelving rock formation at the bottom of the abyss, shoots out rocket-like columns plainly seen from the ledge above. A quarter of a mile below, the river takes another leap of 360 feet, called the Lower or Great

UPPER FALLS.

Falls. The river between, while seemingly placid, from points of observation most readily accessible, and therefore usually visited, is exceedingly rapid, though its remarkable clearness affords a view of its apparently smooth, rocky bottom the entire distance. A footpath leads to the bottom of the Upper Falls, where very fine brook trout fishing may be enjoyed, and midway between this point and the Lower Falls, Cascade Creek enters the river.

Cascade Falls are below the bridge which spans the creek. Their aggregate fall, including the cascades above, is about 130 feet, and a ladder to Grotto Pool allows an inspection of them, though these minor attractions possess little, if any, charm to the sightseer when so near a sight justly rated among the grandest among earth's many marvelously grand scenic displays

RAPIDS ABOVE THE UPPER FALLS.

GREAT FALLS AND GRAND CANYON OF THE YELLOWSTONE.

A short distance beyond Cascade Creek the road passes a point from which the first glimpse of the Grand Canyon is obtained. Inspiration Point may be seen some three miles away. The eye follows the river's course to the brink of the Great Falls, when it suddenly disappears, to be seen again some distance below, meandering, like a slender ribbon of silver, between frowning canyon walls. Near Cascade Creek Bridge a signboard points out the trail to the brink of the Lower Falls, following which (on foot) the visitor soon stands upon a natural platform of rock upon the very edge of the canyon, overlooking the awful plunge of seething waters. At this point, the river, though some 250 feet in width a short distance above, narrows to just seventy-four feet, and while the view is grand almost beyond expression, it is not the best to be obtained of the falls, Point Lookout and Red Rock being regarded the best points from which to see them; however, probably no better view (certainly none more comprehensive) of the canyon is obtainable than that to be had from the platform overlooking the brink of the falls. Gazing down the canyon, Point Lookout is seen rising from 1,200 to 1,500 feet above the river. Almost directly opposite, on the right hand side of the canyon, is Artists' Point, so called from being the position selected by Mr. Thomas Moran from which to paint his celebrated picture now hanging in the nation's capitol at Washington.

GREAT FALLS OF THE YELLOWSTONE.

As seen from "Uncle Tom's" Trail to the bottom of the Canyon.

Grand Canyon **Hotel** is situated on an eminence one thousand feet above the Lower Falls. The view of the surrounding country is especially fine; one overlooks the canyon's gorge, which can only be seen to advantage from the road along the verge and from the several points of observation. Those having extra time to spend in the Park will find this locality filled with interesting features. Aside from the Grand Canyon, the Upper and Lower Falls, a side trip can be made to Mount Washburn and other places in the vicinity. If one has but a day extra, above the regular tour, to stay in the Park, the Grand Canyon is generally the place selected. The hotel is modern throughout and equal to any of the excellent stopping places provided for visitors.

CANYON HOTEL.

CANYON, FALLS AND POINT LOOKOUT.

Point Lookout.—The driveway follows, as nearly as practicable, the very edge of the canyon from the falls to Inspiration Point, about three miles. Point Lookout is about half a mile below the falls, and commands altogether the best combined view of the Great Falls and Grand Canyon. It is fully 1,200 feet above the river and nearer the hotel than any of the several points of observation. Red Rock under Point Lookout, to which a perfectly safe trail leads down the ravine near the point, affords the best view of the falls themselves possible for tourists to obtain.

Grand View.—There are many projections between Lookout and Inspiration, from which glimpses of the canyon may be had. Grand View is about midway between Point Lookout and Inspiration Point, nearly opposite Artists' Point on the opposite side of the canyon. It affords an excellent view of the canyon and of the rugged cliffs about Inspiration Point.

Inspiration Point is considered by many as being, of all points, the best from which to see and appreciate the vast immensity of the canyon, and, although it is two miles from Point Lookout, the grandeur of the view, when considered together with the various other points and projections from which a more or less extended inspection of the canyon may be made, well repays one for the extra effort required.

Inspiration Point is 1,500 feet above the river, and would afford an excellent distant view of both canyon and falls, were it not for Point Lookout intervening, which obstructs a large portion of the latter.

Looking down stream, the view of the canyon is especially fine; though the brilliant coloring of its walls is not so noticeable as above the point. Beside the road, a short distance from this locality, may be seen a large boulder of granite, a most interesting relic of glacial deposit, said by geologists to have been stranded here during the "ice period."

The opposite side of the canyon possesses many excellent points from which to observe both canyon and falls, giving the reverse effects of lights and shadows, which in itself is highly interesting. Particularly is this true of Artists' Point, from which an unobstructed view of both canyon and falls may be had, and is now rendered accessible by a bridge over the Yellowstone above the Upper Falls, which enables one to make a satisfactory inspection of these numerous attractions; among which an eagle's nest, situated upon an inaccessible crag, fully compensates for the trouble of bringing a fieldglass, in the interesting study it affords the beholder. With respect to scenic effects, that obtained in the morning from Inspiration Point and that in the afternoon from the brink of the Great Falls, are considered by many the best. However, each and every hour produces an effect of light and shade possible for no artist to portray.

Believing that the purposes of a Guide Book are best subserved by confining its scope to plain descriptive statements, calculated to enable the reader to readily find, and recognize when found, the subjects concerning which it treats, all attempts at "word pictures" have been rigidly excluded from this little hand-book.

DOWN FROM INSPIRATION POINT.

It would fail, however, to even faintly convey to the mind of the intending visitor, for whose benefit, of course, special reference was had in its inception, anything like an intelligent idea of the wonders of Yellowstone Park, if it neglected to acquaint him with the expressed opinions of some among the many distinguished literateurs, scientists, artists, and others who have carefully inspected this region.

The following interestingly graphic, and, withal, faithful, pen picture of the Grand Canyon and Great Falls of the Yellowstone, by the Rev. Dr. Wayland Hoyt, is subjoined for this purpose:

"Well, we have reached Cascade Creek at last, and a beautiful grove of pine trees, beneath whose shade a clear stream, whose waters are free from the nauseous taste of alkali, furnished a delightful place to camp. Now, dismounting and seeing your horse is well cared for, while the men are unloading the pack-mules and pitching the tents, walk up that trail winding up that hillside, follow it for a little among the solemn pines, and then pass out from the tree shadows and take your stand upon that jutting rock, clinging to it well meanwhile and being very sure of your footing, for your head will surely grow dizzy, and there opens before you one of the most stupendous scenes of nature, the Lower Falls and the awful Canyon of the Yellowstone. And now, where shall I begin, and how shall I, in anywise, describe this tremendous sight; its overpowering grandeur, and, at the same time, its inexpressible beauty?

"Look yonder! Those are the Lower Falls of the Yellowstone. They are not the grandest in the world, but there are none more beautiful. There is not the breadth and dash of Niagara, nor is there the enormous depth of leap of some of the waterfalls of the Yosemite. But here is majesty of its own kind, and beauty, too. On either side are vast pinnacles of sculptured rock. There, where the rock opens for the river, its waters are compressed from a width of 200 feet between the Upper and Lower Falls, to less than 100 feet when it takes the plunge. The shelf of rock over which it leaps is absolutely level. The water seems to wait a moment on its verge;. then it passes, with a single bound, 360 feet into the gorge below. It is a sheer, unbroken, compact, shining mass of silver foam. But your eyes are all the while distracted from the fall itself, great and beautiful as it is, to its marvelous setting; to the surprising, overmastering canyon into which the river leaps, and through which it flows, dwindling to but a foamy ribbon there in its appalling depths. As you cling here to this jutting rock, the falls are already many hundred feet below you. The falls unroll their whiteness down amid the canyon glooms. * * * These rocky sides are almost perpendicular; indeed, in many places the boiling springs have gouged them out so as to leave overhanging cliffs and tables at the top. Take a stone and throw it over; you have to wait long before you hear it strike. Nothing more awful have I ever seen than the yawning of that chasm. And the stillness, solemn as midnight, profound as death. The water dashing there, as in a kind of agony, against these rocks, you cannot hear. The mighty distance lays the finger of silence on its white lips. You are oppressed with a sense of danger. It is as though the vastness would soon force you from the rock to which you cling. The silence, the sheer depth, the gloom burden you. It is a relief to feel the firm earth beneath your feet again, as you carefully crawl back from your perching place.

"But this is not all, nor is the half yet told. As soon as you can stand it, go out on that jutting rock again and mark the sculpturing of God upon those vast and solemn walls. By dash of wind and wave, by forces of the frost, by file of snow plunge and glacier and mountain torrents, by the hot breath of boiling springs, those walls have been cut into the most various and surprising shapes. I have seen the 'middle age' castles along the Rhine; there those castles are reproduced exactly. I have seen the soaring summit of the great cathedral spires in the country beyond the sea; there they stand in prototype, only loftier and more sublime.

"And then, of course, and almost beyond all else, you are fascinated by the magnificence and utter opulence of color. Those are not simple gray and hoary depths, and reaches and domes and pinnacles of sullen rock. The whole gorge flames. It is as though rainbows had fallen out of the sky and hung themselves there like glorious banners. The underlying color is the clearest yellow; this flushes onward into orange. Down at the base the deepest mosses unroll their draperies of the most vivid green; browns, sweet and soft, do their blending; white rocks stand spectral; turrets of rock shoot up as crimson as though they were drenched through with blood. It is a wilderness of color. It is impossible that even the pencil of an artist can tell it. What you would call, accustomed to the softer tints of nature, a great exaggeration, would be the utmost tameness compared with the reality. It is as if the most glorious sunset you ever saw had been caught and held upon that resplendent, awful gorge.

"Through nearly all the hours of that afternoon until the sunset shadows came, and afterwards amid the moonbeams, I waited there, clinging to that rock, jutting out into that overpowering, gorgeous chasm. I was appalled and fascinated, afraid, and yet compelled to cling there. It was an epoch in my life."

Grand Canyon to Norris Geyser Basin.—The twelve-mile drive between the Canyon and Norris is through an undulating pine forest the greater part of the distance. It passes over the "divide" separating the Yellowstone from the Missouri at an altitude of more than 8,000 feet. Between the five and six-mile posts, on the south side of the road is seen the "Twin Trees." Virginia Cascades, about three miles from Norris, are much admired, being quite unlike the many falls and cascades seen throughout the Park.

Mount Washburn.—Until the new road from the Canyon to Mammoth Hot Springs via Mt. Washburn and Tower Falls is open to traffic, a side trip to Mt. Washburn will be made from the Grand Canyon. The new road follows up the valley of Cascade Creek and passes over the mountain through Dunraven Pass. The road to the summit diverges from the main road near the pass to the very top of the mountain, passing over the same and intersecting the main driveway on its northern slope. Especially constructed vehicles, built expressly for the purpose, will be used to transport passengers over the summit. No other vehicle will be allowed on the summit road (saddle animals excepted) owing to its construction and the many places that are exposed to the high winds that at times prevail on the mountain top. Mt. Washburn has an elevation of 10,388 feet above sea level; it's the highest mountain in the reservation. On a clear day, the view from the summit is unsurpassed. Mountain peaks and ranges not visible from any other point on the tour stand out in bold relief.

As has been said, Mt. Washburn is the observatory of the Park; and, while a trip to and from its summit consumes nearly a day, the grand view obtained amply repays the effort.

Dr. Hoyt thus describes it:

"Let us take our stand for a little now upon Mt. Washburn. Its rounded crest is more than 10,000 feet above the level of the sea; and perhaps 5,000 feet above the level of the valley, out of which it springs. Its smooth slopes are easy of ascent. You need not dismount from your horse to gain its summit. Standing there you look down upon the whole grand panorama as does the eagle yonder, holding himself aloft upon almost motionless wings. I doubt if there is another view at once so majestic and so beautiful in the whole world. Your vision darts through the spaces for 150 miles on some sides. You are standing upon a mountain lifting itself out of a vast saucer-shaped depression. Away yonder, where the sky seems to meet the earth on every side around the whole circumference of your sight, are lines and ranges of snow-capped peaks shutting your glances in. Yonder shoots upward the serrated peak of Pilot Mountain, in the Clark Fork range. Joined to that, sweep on around you, in the dim distance, the snowy lines of the Madison range. Yonder join hands with these, the Stinking Water Mountains, and so on and on and around.

"Take now a closer view for a moment. Mark the lower hills folded in their thick draperies of pine and spruce, like dark green velvet of the softest and deepest; notice, too, those beautiful park-like spaces where the trees refuse to grow, and where the prairie spreads its smooth sward freely towards the sunlight. And those spots of steam breaking into the vision every now and then, and floating off like the whitest clouds that ever graced the summer sky, those are the signals of the geysers at their strange duty, yonder in the geyser basins thirty miles away. And those bits of silver, flashing hither and thither on the hillside, amid the dense green of the forests, these are waterfalls and fragments of ice glaciers, which for ages have been at their duty of sculpturing these mountains, and have not yet completed it. And that lovely deep blue sheet of water, of such a dainty shape, running its arms out toward the hills, and bearing on its serene bosom emeralds of islands; that is the sweetest sheet of water in the world; that is the Yellowstone Lake. And that exquisite broad sheen of silver, winding through the green of the trees and the brown of the prairie; that is the Yellowstone River starting on its wonderful journey to the Missouri, and thence downward to the gulf, between 6,000 and 7,000 miles away. But, nearer to us, almost at our feet, as we trace this broad line of silver, the eye encounters a frightful chasm, as if the earth had suddenly sunk away; and into its gloomy depths the brightness and beauty of the shining river leaps, and is thenceforth lost altogether to the view. That is the tremendous canyon, or gorge, of the Yellowstone."

Tower Falls, Yancey's and Petrified Trees.—It is about ten miles from Mt. Washburn to Tower Falls, and three miles from the falls to Yancey's, where the wagon-road is intersected leading from Mammoth Hot Springs to Cook City (a mining camp just outside the northeast corner of the Park). Tower Creek empties into the Yellowstone a few hundred feet below the falls; these are 110 feet high, deriving their name from the peculiar rock formation which rises, tower-like, several hundred feet above their brink. The canyon wall of the Yellowstone, opposite the mouth of Tower Creek, is capped with a layer of lava quite unlike other portions of the canyon.

The petrified trees are one and a half miles from Yancey's and are reached by an easy trail. They are the only specimens of petrified trees, standing in their natural position, to be found in the Park. On Specimen Ridge, across the Yellowstone some ten miles from Yancey's, are numerous specimens of petrifactions, many of them being four and five feet in diameter. Yancey's place is 18 miles from Mammoth Hot Springs; the road over Mt. Washburn will connect with the present road near this point. The best of trout fishing will be found in the Yellowstone near Yancey's, in the East Fork and Slough Creek.

A PETRIFIED TREE.

SIDE TRIPS FROM THE MAMMOTH HOT SPRINGS.

Middle Gardiner Falls and Bunsen Peak.—Surrounding the Mammoth Hot Springs, and not on the main traveled roads, are many interesting places that can be visited by ladies and gentlemen fond of horseback riding. The road to the Middle Gardiner Falls leaves the Golden Gate road about half a mile south from the hotel, passing along the west side of the government inclosure to the West Gardiner River, crosses the same and passes over the eastern slope of Bunsen Peak to Observation Point, which is 1,200 feet above the Middle Gardiner. The walls of the canyon are nearly perpendicular, especially on the east side, and resemble more nearly the Grand Canyon of the Yellowstone than any other place in the Park. The total drop of the falls is 300 feet; the road continues around the slope of Bunsen Peak, intersecting the main road at Golden Gate. The comparatively easy drive around the mountain, though slightly longer, is preferable to returning over the road just followed. The ascent of Bunsen Peak is easily made from the south side; the view from its summit is grand and impressive. To the south the lofty Teton range, though 100 miles distant, can be seen on a clear day; the Gallatin range lies to the west; while Electric Peak, Hot Spring Valley and the Yellowstone range occupy the north and east. No better point can be found in this portion of the Park from which to study the geography of the reserve. From the west entrance of Golden Gate, one has the choice of two routes to the hotel—the regular carriage road and the old road. The latter leads north along the east side of Swan Lake Valley, crossing Terrace Mountain through a pass called Snow Gate; thence down over the formation to the hotel. The entire distance of the trip is about eight miles, and can be easily accomplished in half a day. If the ascent of Bunsen Peak is made, it will be necessary to start in the morning and lunch at Gardiner Falls.

Mount Evarts and East Gardiner Falls.—The gigantic wall, facing the Mammoth Hot Springs and rising some 2,000 feet above Gardiner River, is the west slope of a broad, triangular mountain, comprising an area of twenty square miles, known as Mt. Evarts—a designation that has been attached to it since 1870. The story of Mr. Evarts' adventure and suffering is well known, and the fact that his rescue from a horrible death took place in a little valley just back of the summit of the mountain gives a tinge of romance to the locality and makes the name more than usually appropriate. Mr. Evarts was a member of the Washburn-Langford party who explored the Park in 1870. While this party was in camp at Yellowstone Lake, some fifty miles distant, Mr. Evarts decided to part company with the explorers and return to Bozeman. Being positive that he could reach his destination without the assistance of a guide, he was allowed to depart, mounted on a horse, with a pack horse to carry his provisions and camp equipage. Overtaken on the way by a severe storm, he became bewildered; and while searching on foot for evidences of a trail, lost his eye-glasses, and was unable to return to his horses. For three weeks he wandered helplessly over the country, but was found at last by Scout Jack Barronette upon this mountain in a very precarious condition. Mr. Evarts' rescue was largely due to the prompt investigation set on foot by Gen. Washburn upon the return of his horses to the explorers' camp and, happily, he fully recovered from his thrilling experience and privation.

The stage road from Mammoth Hot Springs to Yancey's passes East Gardiner Falls, which are nearly on a level with the eastern slope of Mt. Evarts; it is possible to drive to the edge of this moutain, which overlooks the Hot Springs Valley and commands one of the finest views of this locality. Mt. Evarts is a favorite summer range for game; and it is not uncommon for visitors to see elk, antelope and mountain sheep while on the mountain. East Gardiner Falls are composed of two cascades, the upper one having a drop of nearly fifty feet, while the lower falls are more broken and have a total drop of about eighty feet; they are but a short distance apart. An excellent view of Gardiner Canyon and the hot springs in the distance may be had from the ledge near the Lower Falls. The East Gardiner Falls are five miles from the hotel; and the summit of Mt. Evarts, overlooking the hot springs, is a mile and a half further. Persons on horseback can cross over Mt. Evarts and strike the Yellowstone trail, which intersects the road from Cinnabar to the hot springs near Gardiner City. Native brook trout fishing is good in Lava Creek, the main branch of the East Gardiner, a mile or more above the falls; and in Black-Tail-Deer Creek, some two miles beyond, the latter being a tributary of the Yellowstone.

THE PARK IN MIDWINTER.

The first attempt to explore this region in the winter occurred in January, 1887. The expedition was headed by Lieutenant Frederick Schwatka, of Arctic fame, and accompanying him were several eastern gentlemen and F. Jay Haynes, as photographer, together with a corps of guides, packers and assistants. The party was outfitted with Arctic "sleeping bags," the Norwegian "ski," the Canadian "web" snowshoe, and toboggans to carry supplies, photographic equipment and astronomical instruments, it being the intention to camp wherever night overtook the party, regardless of the hotels. The expedition consumed three days in reaching Norris, leaving the Mammoth Hot Springs January 2d, and camping at Indian Creek the first night, with the thermometer thirty-seven degrees below zero. The second camp was near Obsidian Cliff. This very slow rate of traveling was due, in a great measure, to the depth and lightness of the snow, in which the toboggans sank readily, making them difficult to draw. At Norris, Lieutenant Schwatka unfortunately fell ill, and was compelled to abandon further exploration.

THE SCHWATKA PARTY PASSING OBSIDIAN CLIFF.

Mr. Haynes, who was specially desirous of obtaining a line of photographic negatives of winter scenes that would embrace the most interesting portions of the great reserve, employed two of the sturdiest men in the Schwatka party, and, accompanied by Edward Wilson, a government scout, pushed on, and succeeded in making a complete circuit of the Park, visiting the Lower and Upper Geyser Basins, the Falls and Grand Canyon, and crossing over Mt. Washburn via Yancey's to the Mammoth Hot Springs.

The fallacy of attempting to drag toboggans was proven in getting to Norris, hence this party resorted to the customary fashion of packing upon their backs their equipment, sleeping bags and provisions, each carrying from thirty to forty-five pounds. The Norwegian snowshoe, or "ski," is a slender runner of tough, springy wood, slightly turned up at the forward end, some four inches wide by twelve feet long, and fitted with a looped thong or strap, into which to insert the foot. They are slid over the snow, the operator carrying a pole some eight feet long to assist him in maintaining his balance, and to be used as a brake when descending mountain sides, without which a velocity would be attained that would be extremely dangerous. In ascending, a "tacking" process is resorted to, and for abrupt places the "corduroy step," which is simply stepping sidewise. The first grand sight presented was in Norris Basin; the great amount of steam congealed on the foliage in the vicinity produced all the fantastic shapes and forms possible to imagine, while the numerous vents sending up their columns of steam resembled a vast manufacturing center.

The telegraph wires, happening to be over a hot spring, were generally broken down by the immense weight of accumulated ice, frequently assuming a diameter of two or three inches. The Upper Basin, however, presented the most striking appearance; the greater quantity of steam and more numerous active geysers presented an increased variety of peculiar effects. During eruptions of Old Faithful and other large geysers in the Upper Basin, the column of steam would rise fully two thousand feet and could be seen from the Lower Basin, ten miles away. The great fall of snow throughout the Park, fully eight feet in depth, gave quite a different aspect to the country; familiar trees, to one visiting the Park in summer, appeared like mere bushes; and fallen timber, sage brush and boulders were entirely hidden. The Grand Canyon was completely changed, its beautifully colored walls were masses of pure white. The Great Falls was a strange sight; the north half was frozen, immense icicles 200 feet in length hanging therefrom; an ice-bridge fully 100 feet high was formed at the base, coming up fully to the spray line, which is usually one-third the height of the falls, and the brink was frozen over, being hidden by an arch of ice fully a dozen feet thick. The trip over Mt. Washburn, in which the entire party nearly lost their lives, was one of hardship and privation, a blinding snowstorm being encountered on the mountain, lasting for three days, in which this little party of four wandered day and night, without food, shelter or fire—an adventurous experience few care to undergo. On the exposed ridges of Mt. Washburn thousands of elk were seen, this being their winter range. The extreme rigors of this section prevent it ever becoming a winter resort.

The circuit of the Park on snowshoes covered nearly 200 miles, the temperature varying from not warmer than ten degrees below zero to fifty-two degrees below during the entire twenty-nine days consumed by this expedition.

GAME OF THE PARK IN WINTER.

Early in March, 1894, a party was organized at Ft. Yellowstone for the purpose of visiting the winter ranges of the game, to ascertain, as near as possible, the exact number of buffalo that still exists, and secure photographs of the same. The party consisted of Captain Scott, Lieut. Forsyth, Scout Burgess, Mr. Burns, Photographer Haynes and three non-commissioned officers. Mounted on the Norwegian snowshoe, with packs of sleeping bags, provisions and camera, they proceeded directly to Hayden Valley via Norris and the Grand Canyon. As most of the buffalo congregate there during the winter months, they found eighty-one buffalo in the valley, seventy-three comprising the main herd, and numerous small groups of elk aggregating fully 300. After a stay of several days in Hayden Valley the party went to Yellowstone Lake. Cap-

ELK, HAYDEN VALLEY.

tain Anderson, superintendent of the Park, had instructed Scout Burgess not to overlook the country east of the lake, as a small herd of buffalo usually winter there. The first day out from the lake only elk were seen by the scout and his companion, there being no sign of buffalo. They went into camp about twelve miles up Pelican Creek.

The second day they discovered, in a secluded spot, the "cache" of a poacher, very much to their surprise, as it was supposed that no one was in the Park killing game. The "cache" consisted of a canvas tepee, sleeping bag, provisions and toboggan, and six buffalo heads suspended in a tree near by. A trace of fire in the tepee led the scout to believe that the poacher was in the vicinity, and the next move was to capture him. It had been snowing constantly all the morning, and all snowshoe tracks leading from the camp were entirely obliterated. Some five miles from the camp they heard five or six rifle shots fired in rapid succession. Hastening through the timber to the opening in the direction of the firing, they came directly upon the poacher. He had driven six of the buffalo into the deep snow and slaughtered the entire band. Knowing these men to be of a desperate character, and being armed only with a pistol, it was a brave act for Scout Burgess to arrest him. Fortunately it was snowing hard, and the approach of the scout was not noticed by the poacher or his faithful dog until the arrest was made. He was taken to the Lake Hotel and escorted from there to the guard house at Ft. Yellowstone. Besides the twelve buffalo that were killed by this poacher, a small herd of seven were seen in the Pelican country, making less than 100 now in existence. If these can be protected they will increase rapidly, otherwise the only remaining species of large American game (the bison) will soon be exterminated. Elk were seen in the foothills of Mt. Washburn, on Specimen Ridge, along the east fork of the Yellowstone, on Slough Creek and

BUFFALO, HAYDEN VALLEY.

along the Yellowstone to Mt. Evarts, in great numbers. Fully 5,000 wintered in the above localities. Small bands of mountain sheep, deer and antelope were seen on Mt. Evarts. The open water of the Yellowstone between the lake and falls was alive with duck and swan. The red fox and coyote were numerous, and an occasional black fox and footprints of mountain lion and bear were seen. The party was in the Park about thirty days and traveled over 300 miles.

History and Early Explorations.—The following brief history of the Park and account of the early exploration of the region is taken from the report made to the late Dr. F. V. Hayden, chief of Geological Survey of Territories, by Henry Gannet, E. M., on the geographical field work of the U. S. Geological Survey during the season of 1878:

"The first authentic information regarding the great natural wonders of the Park was derived from a prospecting party under the leadership of Captain W. W. DeLacy, who, in 1863, visited the Lower Geyser Basin. Previous to this time it seems that the region was known to but a few hunters and trappers, and their tales were treated as the wildest of romancing, as, indeed, many of them were, the mind of the trapper being singularly prone to exaggeration. The earliest reference to the hot springs is in stories of a trapper by the name of Colter (or Coulter) who accompanied Lewis and Clark's celebrated expedition across the continent. On the return of this expedition, when below the mouth of the Yellowstone, Colter was discharged, at his own request, and immediately returned to the country above the forks of the Missouri. In this neigborhood, probably in the Jefferson, his companion, Potts, was killed by Blackfeet Indians, and he was captured. Almost miraculously he escaped from them, and, entirely naked, made his way to a trading post on the Big Horn. After this he lived for a year or more among the Bannacks, whose range included what is now the Yellowstone Park. Either during his perilous journey after his escape from the Blackfeet, or during his sojourn among the Bannacks, he became acquainted with the region of the hot springs and geysers, for we find him in Missouri in 1810, telling marvelous tales of lakes of burning pitch, of land on fire, hot springs and geysers. His stories were, of course, treated as travelers' tales, and 'Coulter's Hell' was classed with Lilliput, Symmes' Hole, and other inventions of over-developed imaginations.

"Later we find the knowledge of this country more generally diffused among this people. Colonel Reynolds, in his report on the 'Exploration of the Yellowstone,' in 1859-60, refers to 'some of their Munchausen tales,' as follows:

"One was to this effect: 'In many parts of the country petrifications and fossils are very numerous, and, as a consequence, it was claimed that in some locality (I was not able to fix it definitely) a large tract of sage is perfectly petrified, with all the leaves and branches in perfect condition, the general appearance of the plain being like that of the rest of the country, but all is stone; while the rabbits, sage hens and other animals usually found in such localities are still there, perfectly petrified, and as natural as when they were living; and, more wonderful still, the petrified bushes bear the most wonderful fruit; diamonds, rubies, sapphires, emeralds, etc., etc., as large as black walnuts, are found in abundance.'

"This story, absurd as it sounds, has a large basis in fact. The narrator, however, had mixed up distinct phenomena and over all had spread lavishly the coloring of his imagination. There are fields of sage, as well as bits of forest, which, lying in immediate proximity to groups of springs, have been petrified while standing. The hot, silicious water from the springs is drawn up through the pores of the bark by capillary attraction, and, depositing silica wherever it goes, the tree or bush is rapidly transformed into rock.

"The story of the remarkable fruit borne by these stone trees is not far from correct, the main difference between the story and the fact being that the fruit is borne on the outside and inside of the trunk of the trees, instead of on the ends of the branches. The mineral species are not as given in the story, either, but this is a matter of no vital importance. In the process of the silicification of wood the last result of all is the production of quartz crystals. The trunk is converted totally into crystalline quartz, radiating from within outward, the crystals being all crowded out of shape. The inside and outside of the hollow cylinder of quartz, which represents the former tree, are covered with the characteristic quartz pyramids. Such products of silicification are very abundant in the Park, particularly on Amethyst Ridge, and are, undoubtedly, the stone fruit of the petrified trees and bushes. The crystals are colorless, amethystine or yellow, and, according to the color, are known to the mountain man as diamonds, amethyst, topaz, etc. It is unnecessary to say that the part of the story relating to animal life was manufactured from the whole cloth.

"Many other legends had long been current among mountain men, some of which are briefly referred to in Colonel Norris' report to the Secretary of the Interior for 1878, but none of them seem to have attracted any attention. That white men have been in the Park prior to any printed records is evidenced by the discovery by Colonel Norris, as noted in his report above referred to, of a block house near the Grand Canyon, of a cache of martin traps near Obsidian Canyon, and other relics of the early trappers.

"In 1863, Captain W. W. DeLacy, in command of a large party of prospectors, left Montana to prospect on the upper waters of the Snake. Striking that river near the junction of Henry's Fork, they followed up the main river through the canyon, prospected in Jackson's Hole, and, not finding gold in paying quantities, they broke up the party, some returning one way, some another. Captain DeLacy, with a portion of the party, followed up the Snake and Lewis Fork, discovering Lewis and Shoshone (DeLacy's) Lakes, the Shoshone and the Lower Basins. The geographical work done by Captain DeLacy on this trip was embodied in a map of Montana, drawn by himself, and published by authority of the territory in 1864-65, and the material thus made public was afterwards used by the land office in the compilation of maps of that region.

"The results of this trip seem to have attracted little or no attention, for we hear of no one going into the country until 1869, when the two prospectors, Cook and Folsom, made a prospecting tour through the Park. They followed the Yellowstone up to the mouth of the East Fork, then up the latter stream for a few miles, crossing over to the Yellowstone at the Great Falls; thence they went up this stream to the foot of the lake, and around the east side of the latter to the extremity of the west arm; thence crossing over to Shoshone Lake and Lower Geyser Basin on the Madison or Firehole, and finally left the country by following down the Madison River.

"Their story, written by Mr. David E. Folsom, and published in the Chicago Western Monthly for July, 1870, immediately attracted attention, and the following summer a large party, composed of prominent citizens of Montana, under the leadership of General Washburn, then Surveyor General of Montana, was made up for the purpose of exploring this region. Among the party was Hon. N. P. Langford, first superintendent of the Park; Hon. Cornelius Hedges, who first proposed setting apart this region as a National Park; Hon. J. C. Evarts and S. T. Hauser, accompanied by a small escort from Fort Ellis in charge of Lieut. G. C. Doane.

"This party made quite extensive explorations on the Yellowstone and Madison Rivers. Passing up the Yellowstone by the well-known trail, they traveled completely around the lake, visiting all localities of interest along the route, with the single exception of Mammoth Hot Springs, on Gardiner River.

"The following year, 1871, Captains Barlow and Heep, U. S. A., made a reconnaissance of this country, and presented the results in a brief report and a map of their route.

"In the same year Dr. Hayden devoted a portion of the season to a reconnaissance of this region, making quite an extended tour through it. The result of this work, including geological reports, maps, etc., was published in the annual report for that year. This was sufficient to fix the public attention upon this great collection of natural wonders, and when Dr. Hayden presented to Congress a proposition to reserve this section from settlement as a National Park, it was adopted with little opposition. The following year, 1872, Dr. Hayden continued the reconnaissance of the Park, and the country to the north and south of it, publishing the results in the reports of that year and in a series of maps.

"In 1873, Captain W. A. Jones, U. S. A., took a large party through it. He entered it from the head of the Stinking Water, crossing one of the many passes near Mt. Chittenden. After visiting most of the points of interest in the Park, he went out via the Upper Yellowstone, on the way verifying the old trappers' legend about the 'Two Ocean River,' and discovering a practical pass (Togwotee Pass) and route from the south to the Park. This discovery was by far the most valuable result of the expedition.

"In 1872, Captain William Ludlow, U. S. A., in

charge of a reconnaissance in Central Montana, made a flying trip to the Park. He developed little that was new save accurate measurements of the Upper and Lower Falls of the Yellowstone."

THE ACT OF DEDICATION.

Be it enacted by the Senate and House of Representatives of the United States of America in Congress assembled:

That the tract of land in the Territories of Montana and Wyoming, lying near the headwaters of the Yellowstone River, and described as follows, to-wit: Commencing at the junction of Gardiner River with the Yellowstone River, and running east to the meridian passing ten miles to the eastward of the most eastern point of Yellowstone Lake; thence south along the said meridian to the parallel of latitude passing ten miles south of the most southern point of Yellowstone Lake; thence west along said parallel to the meridian passing fifteen miles west of the most northern point of Madison Lake; thence north along said meridian to the latitude of the junction of the Yellowstone and Gardiner Rivers; thence east to place of beginning,—is hereby reserved and withdrawn from settlement, occupancy or sale under the laws of the United States, and dedicated and set apart as a public park or pleasure ground for the benefit and enjoyment of the people; and all persons who shall locate, settle upon or occupy the same or any part thereof, except as hereinafter provided, shall be considered trespassers and removed therefrom.

Sec. 2. The said public park shall be under the exclusive control of the Secretary of the Interior, whose duty it shall be, as soon as practicable, to make and publish such rules and regulations as he may deem necessary and proper for the care and management of the same. Such regulations shall provide for the preservation from injury or spoliation of all timber, mineral deposits, natural curiosities or wonders within said park and their retention in their natural condition.

The Secretary may, in his discretion, grant leases for building purposes, for terms not exceeding ten years, of small parcels of ground, at such places in said park as shall require the erection of buildings for the accommodation of visitors; all the proceeds of said leases, and all other revenues that may be derived from any source connected with said park, to be expended under his direction, in the management of the same, and the construction of roads and bridle paths therein. He shall provide against the wanton destruction of the fish and game found within said park, and against their capture or destruction for the purpose of merchandise or profit. He shall also cause all persons trespassing upon the same after the passage of this act to be removed therefrom, and generally shall be authorized to take all such measures as shall be necessary or proper to fully carry out the objects and purpose of this act.

RULES AND REGULATIONS OF THE YELLOWSTONE NATIONAL PARK.

The following and eminently proper rules have been prescribed for the government of the Park and the protection of its multifarious objects of public interest and noble game:

1. It is forbidden to remove or injure the sediments or incrustations around the geysers, hot springs or steam vents; or to deface the same by written inscription or otherwise; or to throw any substance into the springs or geyser vents; or to injure or disturb, in any manner, or to carry off any of the mineral deposits, specimens, natural curiosities or wonders within the Park.

2. It is forbidden to ride or drive upon any of the geyser or hot spring formations, or to turn loose stock to graze in their vicinity.

3. It is forbidden to cut or injure any growing timber. Camping parties will be allowed to use dead or fallen timber for fuel.

4. Fires shall be lighted only when necessary, and completely extinguished when not longer required. The utmost care should be exercised at all times to avoid setting fire to the timber and grass, and any one failing to comply therewith shall be peremptorily removed from the Park.

5. Hunting or killing, wounding or capturing of any bird or wild animals, except dangerous animals, when necessary to prevent them from destroying life or

inflicting an injury, is prohibited. The outfits, including guns, traps, teams, horses or means of transportation used by persons engaged in hunting, killing, trapping, ensnaring or capturing such birds or wild animals, or in possession of game killed in the Park under other circumstances than prescribed above, will be forfeited to the United States except in cases where it is shown by satisfactory evidence that the outfit is not the property of the person or persons violating this regulation, and the actual owner thereof was not a party to such violation. Firearms will only be permitted in the Park on the written permission of the superintendent thereof. On arrival at the first station of the Park guard, parties having firearms will turn them over to the sergeant in charge of the station, taking his receipt for them. They will be returned to the owners on leaving the Park.

6. Fishing with nets, seines, traps, or by use of drugs or explosives, or in any other way than with hook and line, is prohibited. Fishing for purpose of merchandise or profit is forbidden by law. Fishing may be prohibited by order of the Superintendent of the Park in any of the waters of the Park, or limited therein to any specified season of the year until otherwise ordered by the Secretary of the Interior.

7. No person will be permitted to reside permanently or to engage in any business in the Park without permission, in writing, from the Department of the Interior. The Superintendent may grant authority to competent persons to act as guides, and revoke the same at his discretion, and no pack trains shall be al-

lowed in the Park unless in charge of a duly registered guide.

8. The herding or grazing of loose stock or cattle of any kind within the Park, as well as the driving of such stock or cattle over the roads of the Park, is strictly forbidden, except in such cases where authority is granted by the Secretary of the Interior.

9. No drinking saloon or bar-room will be permitted within the limits of the Park.

10. Private notices or advertisements shall not be posted or displayed within the Park, except such as may be necessary for the convenience or guidance of the public, upon buildings on leased ground.

11. Persons who render themselves obnoxious by disorderly conduct or bad behavior, or who violate any of the foregoing rules, will be summarily removed from the Park, and will not be allowed to return without permission in writing from the Secretary of the Interior or the Superintendent of the Park. Any person who violates any of the foregoing regulations will be deemed guilty of a misdemeanor, and be subjected to a fine, as provided by the Act of Congress, approved May 7, 1894, "to protect the birds and animals in Yellowstone National Park, and to punish crimes in said park, and for other purposes," of not more than one thousand dollars or imprisonment not exceeding two years, or both, and be adjudged to pay all costs of the proceedings.

FAUNA AND FLORA OF THE PARK.

It is but proper that the reader of even a guide-book be given some idea of the animal and plant life to be found in this region; however, it should be borne in mind that an exhaustive treatise on this subject is not attempted herein.

Among the wild animals to be found in the Park are: Buffalo, moose, elk, big horn (mountain sheep), deer, antelope, bear, mountain lion (panther), wolf, fox (red, gray and black), coyote, beaver, otter, mink, marten, sable, muskrat, ermine, rabbit, badger, porcupine, hare, squirrel, chipmunk, ground hog, wolverine and skunk. Among the birds (principally migratory) are: Grouse, owl, hawk, eagle, vulture, duck (in great variety), goose, brant, pelican, swan, crane, crow, raven, magpie, lark, bluejay, blackbird and robin.

Reptiles are rare, though the rattlesnake is found in parts of the Yellowstone Valley, below an altitude of 6,000 feet.

Among the animals enumerated the larger varieties are only occasionally met with, and then, as a general thing, only in the more inaccessible and densely wooded portion of the Park, the latter being about three-fourths its entire area.

The principal varieties of trees found here are: Black spruce, fir (black, red and balsam), white pine, jack pine, red cedar, aspen (poplar), dwarf maple and willow; while among the shrubs may be seen the sage, choke-cherry, gooseberry, bullberry, currant and buffalo-berry. Grasses are quite abundant, the predominating varieties being the buffalo, bunch and gramma, which are wonderfully nutritious. Rye grass and wild timothy are seen in many places.

Wild flowers, of almost every hue, and in well-nigh endless variety, are quite plentiful and exceedingly hardy, often withstanding severe frosts without injury. Singularly, also, the more beautiful varieties are found upon the higher elevations, such as Mt. Washburn, Electric Peak and the like, which at certain seasons are gaily bedecked with flowers of rare color and fragrance; however, many rare specimens of flora are to be found in the lower altitudes.

FISH AND FISHING.

To many, one of the most interesting features of the Park is its excellent fishing.

During the summer of 1889, by order of the U. S. Fish Commissioner, a large number of young trout were placed in Park streams. In connection with the rivers and lakes having native mountain trout, the Park cannot be compared with other places as a resort for the angler. It has over 250 miles of trout streams, and five or six varieties. Native trout (*Salmo myhiss*) are found in Yellowstone River, Gardiner River, Black-Tail-Deer Creek, Lava Creek, East Fork of Yellowstone River, Slough Creek, Yellowstone Lake, Snake

River, Heart Lake and Madison River below the Falls of Firehole; Eastern Brook Trout (*Salvelinus fontinalis*), in Upper Gardiner, Willow Creek, Indian Creek and Glenn Creek; Loch Leven Trout (*Salmo trutta levenensis*), in Firehole River, Nez Perces Creek and the Madison above Keppler Cascades; Rainbow Trout (*Salmo iridens*), Madison River below the Falls of Firehole.

Tourists making the usual tour of the Park will have only one opportunity for first-class fishing unless stop-overs are made. This will be at the lake, in the outlet about a mile from the hotel.

LOCH LEVEN AND RAINBOW TROUT OF THE MADISON.

The fishing grounds in the vicinity of Mammoth Hot Springs Hotel are the Gardiner River, Yellowstone River, Black-Tail-Deer Creek, Lava Creek, Willow Creek and Indian Creek.

Near the Fountain Hotel: Firehole River, Nez Perces Creek and Madison River.

At Yancey's: In Yellowstone River, East Fork and Slough Creek. There is no time during the Park season (June to October) that the fishing is not first-class in some part of the Park. Camping parties will find the region in the vicinity of Yancey's an excellent place during August and September.

HOTELS OF THE PARK.

Are six in number: "Mammoth" at Mammoth Hot Springs, "Norris" at Norris Geyser Basin, "Fountain" at Lower Geyser Basin, "Old Faithful Inn" at Upper Geyser Basin, "Lake" at outlet Yellowstone Lake, and "Canyon" at the Grand Canyon. They are under one management, the Y. P. A. (Yellowstone Park Association), who construct and operate the hotels in the Park. Concessions are granted them by the Honorable Secretary of the Interior, and all rates and charges made to patrons require the approval of the department. The private "Telegraph Line" of the Y. P. A. is connected with the Western Union Telegraph Lines from all the hotels. In furnishings and table service these hostelries compare favorably with those of other resorts. They are electric-lighted and steam-heated. Are open to travelers during the season, June, July, August and September.

CAMPING IN THE YELLOWSTONE.

PARK CAMPING GROUNDS.

Mile posts are lettered as follows:

G. C.—For Gardiner City.
M. S.—For Mammoth Hot Springs.
N. B.—For Norris Geyser Basin.
R. S.—For Riverside Station.
F. H.—For Fountain Hotel.
U. B.—For Upper Geyser Basin.
W. T.—For West Thumb, Yellowstone Lake.
L. H.—For Hotel, Yellowstone Lake.
C. J.—Canyon Junction, near Grand Canyon.

Camping parties should inform themselves as to the rules and regulations of the Park and pay particular attention to extinguishing their camp fires.

It is forbidden to camp within 100 feet of any main traveled road.

The following camping places are used by parties making a tour of the Park:

Camp No. 1, along the Gardiner River between Gardiner City and Mammoth Hot Springs. These camping grounds are used considerably; wood is rather scarce, as well as grass; water plenty and excellent; the best camp for the stock is near the Boiling River about three miles from Gardiner City. It is reached by continuing up the Gardiner River on the road along the east bank which leaves the main road near the post garden; the river can be forded to the camp, while the new road that crosses near the garden ascends the mountain away from the river.

Camp No. 2, is Mammoth Hot Springs Camp, about a mile south of the hotel on the old Golden Gate road, near the one-mile post. An excellent spring near the road; good grass and plenty of timber near at hand. This camp is convenient to the Hot Spring formations.

Camp No. 3, at Golden Gate, three and one-half miles from Mammoth Hot Springs, just through Golden Gate Canyon, either to the south, after crossing Glen Creek, or to the north before crossing. Excellent water, plenty of good grazing and wood quite near. The south camp can be made in a gulch a few hundred yards from the road; it is surrounded by timber and well protected.

Camp No. 4, on Gardiner River, seven miles from Mammoth Hot Springs; good water and grass; wood plenty, and timber for protection; also on Willow Creek.

Camp No. 5, ten and one-quarter miles from Mammoth Hot Springs, near the Apollinaris Spring in the south end of Willow Park. An abundance of wood, water and grass.

Camp No. 6, thirteen and three-quarter miles from Mammoth Hot Springs. A sign board reads: "100 yards to good camp, wood, water and grass."

Camp No. 7, at Norris. Pass the junction of the Fountain and Canyon road, follow the latter across the bridge above one-quarter mile and the best camping ground in this vicinity is found; it is nearer the basin, has excellent water, plenty of wood and grass.

Camp No. 8, Elk Park Camp. In going south from Norris, cross Elk Park, pass the two-mile post about one-quarter of a mile, enter the timber on main road about 100 feet and a road leads to the left into a small park—a well protected camp, wood and grass plenty, water in the Gibbon about 300 yards.

Camp No. 9, in Gibbon Meadow, mouth of Gibbon Canyon, between the three and four-mile posts south of Norris. The favorite camp near the river is west of the road near the four-mile post, plenty of wood in the vicinity, and an abundance of grass and water. The camps east of the road are nearer wood, while water is not as convenient.

Parties Entering at Western Entrance will find excellent camping grounds at the junction of the Gibbon and Madison Rivers.

Camp No. 10, Lower Geyser Basin. There is not any good camping ground between Gibbon Meadow and the valley of the Firehole. Near the seventeen-mile post south of Norris, good camps can be made along the river up to the junction of Nez Perces Creek, along Nez Perces for a mile on the north bank, and on the west bank of the Firehole for a couple of miles in the vicinity of the old Firehole Hotel; also on the east bank of the Firehole about half a mile south of the Cavalry Cantonment. No camping is allowed near the Fountain Hotel, Geyser and Paint Pots.

CAMP No. 11, near Excelsior Geyser on the banks of the Firehole, about half a mile south of the second bridge above Excelsior; plenty of wood and water, with grass on each side of the river, which can be forded near the island.

CAMP No. 12, Biscuit Basin, six and one-half miles from Fountain Hotel on the Firehole River, with plenty of wood and grass. This camp is on the north edge of the Upper Geyser Basin, about two miles from Old Faithful, situated in the south end of the basin.

CAMP No. 13, near Riverside Geyser. Camping is not allowed in the Upper Basin between the Riverside Bridge and Old Faithful. Riverside camp is reached by leaving the main road just before crossing the bridge. Wood and water are near at hand, and the open parks back afford fair grazing for the stock. This is the most central camp in the Upper Basin, being near several of the large geysers.

CAMP No. 14, near Old Faithful. Leave the main road about 200 yards south of Old Faithful; on the banks of the Firehole will be a fair camping place for a medium sized party. Grazing grounds are limited in this section.

CAMP No. 15, at the Lone Star Geyser, three and one-half miles south of the Upper Geyser Basin. Leave the main road at the three-mile post and follow the road leading to the geyser. Excellent water with an abundance of wood and grass.

CAMP No. 16, West DeLacy Creek, seven and three-quarter miles from the Upper Basin on the Pacific side of the continental divide, on the west branch of DeLacy Creek, which empties into Shoshone Lake. Well supplied with wood, water and grass.

CAMP NO. 17, on the east branch of DeLacy Creek, one mile east of Camp 16. Either of these camps are first-class and are the only camping grounds between Lone Star and Yellowstone Lake. Teton Point is between Camp 16 and Camp 17.

CAMP NO. 18, Thumb Bay, Yellowstone Lake. The only available camping grounds here are on the creek south of the lunch station. At times the creek is dry and the only good water is the lake. Plenty of grass is found up the creek—wood in abundance.

CAMP NO. 19, on Yellowstone Lake. The first good camping grounds north of Thumb Bay on the road to the hotel are thirteen and one-half miles, five and one-half from the outlet. Plenty of wood and water and good grazing a short distance back from the lake shore.

CAMP NO. 20, on Bridge Bay near the Natural Bridge, seventeen miles from West Thumb, two miles from hotel. This camp is fully as good as Camp 19 and nearer the lake outlet by three and one-half miles.

CAMP NO. 21, at the outlet of Yellowstone Lake. Excellent camping grounds are found a mile or two north of the hotel. A creek of cold spring water, wood and grass in abundance.

CAMP NO. 22, at Mud Geysers, seven and one-half miles from the lake and eight and one-half from the Canyon. Good camping grounds.

CAMP NO. 23, Grand Canyon Camp. The best camping place at the Canyon is one mile south of Canyon Junction. Many parties camp near the Junction, but the camping grounds one mile south are far better. Wood and grass near at hand, the river for water. No camping is allowed nearer the Canyon than at the Junction. Good camping grounds are found on Cascade Creek near the crossing a mile or two north of the hotel, but not convenient for sightseers.

WYLIE PERMANENT CAMPS.

The Wylie Camping Co. has established throughout the Park a number of permanent camps for the convenience of their patrons. The four night stations are "Willow Park," "Upper Basin," "Yellowstone Lake," and "Grand Canyon." The two lunch stations are "Gibbon Canyon" and "Thumb Bay—Yellowstone Lake." Sleeping accommodations are also furnished at these lunch stations for a limited number, such as bicyclers, etc. In addition to these camps the Wylie Co. provides transportation for the tour of the Park from Gardiner and return. Their regular tour is six days, with the following itinerary: First day, Gardiner to Willow Park; second day, Willow Park to Upper Basin; third day, at Upper Geyser Basin; fourth day, Upper Basin to Yellowstone Lake; fifth day, Yellowstone Lake to the Grand Canyon; sixth day, Grand Canyon to Gardiner. For particulars address Wylie Camping Co., Bozeman, Mont., October until June; Gardiner, Mont., June until October.

JACKSON HOLE HUNTING GROUNDS.

Parties after big game—Elk, Moose, Deer and Mountain Sheep—will find the Jackson Hole section a hunter's paradise. The guides must be residents of Wyoming and each hunter must have a Wyoming license, which allows him to kill but a limited number. There are three routes into Jackson Hole—from Gardiner, Montana, on the Northern Pacific Railway, south through the Park; from St. Anthony, Idaho, on the Oregon Short Line Railroad, east around the Tetons, and from points on the Union Pacific Railroad between Rawlins and Green River.

PARK TRANSPORTATION.

Two authorized stage coach lines operate in connection with the hotels in the Park, one from the Northern entrance, the other from the Western.

NORTHERN ENTRANCE, VIA GARDINER.

The Yellowstone Park Transportation Co. connect with the Northern Pacific Ry. at Gardiner, operating six-horse Concord coaches to Mammoth Hot Springs and for the tour of the Park four-horse Concord coaches. Parties desiring special or private conveyances can arrange with the Transportation Co. at Mammoth Hotel. Saddle and pack animals furnished fishing parties; also outfits for hunting parties going outside the Park. Tourists holding tickets, **Gardiner to O. S. L.**, are transferred to the M-Y coaches at Norris after making the tour of the Park. For particulars address

THE YELLOWSTONE PARK TRANSPORTATION CO.

YELLOWSTONE PARK, June until October

HELENA, MONTANA October until June

WESTERN ENTRANCE, VIA O. S. L.

The Monida & Yellowstone Stage Co. operate a four-horse Concord line from the Oregon Short Line R. R., to and through the Park. Passengers holding tickets, **O. S. L. to Gardiner**, are transferred to the six-horse coaches of the Y. P. Transportation Co. at Mammoth Springs after making the tour of the Park. For particulars and description of the route address

THE MONIDA & YELLOWSTONE STAGE CO.

YELLOWSTONE PARK, June until October.

ST. PAUL, MINNESOTA, October until June.

The Yellowstone Park and How It Was Named.

The Devil was sitting in Hades one day,
In a very disconsolate sort of a way.
One could tell from his vigorous switching of tail,
His scratching his horn with the point of his nail,
That something had gone with His Majesty wrong,
The steam was so thick and the sulphur so strong.
He rose from his throne with a gleam in his eye,
And beckoning an agate-eyed imp standing by,
Commanded forthwith to be sent to him there
Old Charon, employed in collecting the fare
Of the wicked, who crossed the waters of Styx,
And found themselves soon in a deuce of a fix.

Old Charon, thus summoned, came soon to his chief.
As the Devil was angry, the confab was brief.
Says the Devil to Charon, "Now, what shall I do?
The world it grows worse and grows wickeder, too;
What with Portland, Chicago, Francisco, New York,
I get in my mortals too fast for my fork;
I haven't the room in these caverns below,
St. Peter, above, is rejecting them so.
So hie you, my Charon, to earth, far away.
Fly over the globe without any delay,
And find me a spot quite secluded and drear,
Where I can drill holes from the center in here.
I must blast out more space; so survey the spot well,
For the project on hand is the enlargement of Hell.

"But recollect one thing, Old Charon, when you
Can locate the district where I can bore through,
There must be conveniences scattered around
To carry on business when I'm above ground.
An 'ink-pot' must always be ready at hand
To write out the names of the parties I strand.
There must be a 'punch bowl,' a 'frying-pan,' too,
A 'cauldron' in which to concoct a 'ragout.'
An 'old faithful' sentinel showing my power
Must shoot a salute on the earth every hour,
And should any mortal by accident view
The spot you have chosen, why, this you must do:
Develop a series of pools, green and blue,
That while these poor earth bugs may beauties admire
They'll forget that below I'm poking the fire.
Now fly away, Charon, be quick as you can,
For my place here's so full that I can't roast a man.

To earth flew fleet Charon, to regions of ice;
He found these too cold—so away in a trice,
He sought a location in Africa's sands,
He prospected, and finding too much on his hands
He cut out Australia, Siberia, too,
The north part of China—no! they would not do;
Till just as about to relinquish the chase
He stumbled upon a most singular place.
'Twas deep in the midst of a mountainous range,
Surrounded by valleys secluded and strange,
In a country the greatest, the grandest, the best
To be found upon earth—America's West.
Here the crust seemed quite thin and the purified air,
With the chemicals hidden around everywhere,
Would soon make the lakes that the Devil desired;
So he flew to Chicago and there to him wired:
"I've found you a place never looked at before;
You may heat up the rocks, turn on water, and bore."

Then the Devil with mortals kept plying the fire,
Extracting the water around from the mire,
And boring great holes with a terrible dust,
Till soon quite a number appeared near the crust.
Then he turned on the steam—and lo! upward did fly,
Through rents in the surface, the rocks to the sky.
Then with a rumble there came from each spot,
Huge volumes of water remarkably hot,
That had been there in caverns since Lucifer fell—
Thus immensely enlarging the confines of Hell.
And it happens that now when Old Charon brings in
A remarkable load of original sin,
That His Majesty quietly rakes up the coals,
And up spouts the water, in jets, through the holes,
One may tell by the number of spurts when they come,
How many poor mortals the Devil takes home.

But Yankees can sometimes, without doing evil,
O'ermatch in sagacity even the Devil.
For not long ago Uncle Sam came that way
And said to himself, "Here's the Devil to pay.
Successful I've been in all previous wars;
Now Satan shall bow to the Stripes and the Stars.
This property's mine, and I hold it in fee;
And all of this earth shall its majesty see.
The deer and the elk unmolested shall roam,
The bear and the buffalo each have a home;
The eagle shall spring from her eyrie and soar
O'er crags in the canyons where cataracts roar;
The wild fowls shall circle the pools in their flight,
The geysers shall flash in the moonbeams at night,
Now I christen the country—let all nations hark!
I name it the Yellowstone National Park."

WM. TOD HELMUTH.

Grand Canyon, August, 1894.

TABLE OF DISTANCES.

	Miles.
Gardiner to Mammoth Hot Springs	5
Mammoth Hot Springs to Norris	20
Norris to Fountain	20
West Entrance to Fountain	21
Fountain to Upper Basin	10
Upper Basin to West Thumb	19
Thumb to Lake Hotel (Yellowstone)	16
Lake to Canyon Junction	17
Canyon Junction to Norris	11
Grand Canyon to Yancey's	23
Yancey's to Mammoth Hot Springs	18
Entire Circuit of the Park (Stage Route) about	150

ELEVATIONS OF THE PARK.

	Feet.
Mammoth Hot Springs Hotel	6,387
West Entrance Golden Gate	7,300
Norris Geyser Basin	7,527
Lower Geyser Basin	7,252
Upper Geyser Basin	7,394
Yellowstone Lake	7,788
Grand Canyon Hotel	7,710

ALTITUDE OF MOUNTAINS.

	Feet.
Electric Peak	11,155
Quadrant Mountain	10,127
Mt. Evarts	7,600
Bunsen Peak	8,775
Mt. Washburn	10,388
Mt. Langford	10,902
Mt. Sheridan	10,385
Index Peak	11,702
Grand Teton	13,654

Spotted something that isn't quite right?

Let us know and we'll fix it for future reprints.
And as a thank you for your diligence, we'll
send you a free Sothis book of your choice.

publisher@sothis.press

This page intentionally left blank

Made in United States
North Haven, CT
05 November 2026

21817309R00088